Quantum Computation and Quantum Information

A Mathematical Perspective

GRADUATE STUDIES IN MATHEMATICS 243

Quantum Computation and Quantum Information

A Mathematical Perspective

J. M. Landsberg

AMS
AMERICAN MATHEMATICAL SOCIETY
Providence, Rhode Island

2020 *Mathematics Subject Classification.* Primary 81P45, 81P68, 68Q12, 94A15, 20G05, 94A24.

For additional information and updates on this book, visit
www.ams.org/bookpages/gsm-243

Library of Congress Cataloging-in-Publication Data

Cataloging-in-Publication Data has been applied for by the AMS.
See http://www.loc.gov/publish/cip/.
DOI: https://doi.org/10.1090/gsm/243

∞ The paper used in this book is acid-free and falls within the guidelines established to ensure permanence and durability.
Visit the AMS home page at https://www.ams.org/

10 9 8 7 6 5 4 3 2 1 29 28 27 26 25 24

Contents

Preface

This book is for a one-semester course in quantum computing and quantum information theory. It emphasizes the mathematical aspects and the historical continuity of both algorithms and information theory.

Quantum computing and quantum information theory have often been presented as radical departures from classical computation and information theory because of the "strangeness" of quantum mechanics. If one takes a broader view, the picture looks different.

Shor's factoring algorithm is one in a long line of surprising efficient algorithms. This book presents algorithms leading up to Shor's algorithm, e.g., the Schönhage-Strassen polynomial multiplication algorithm and the Miller-Rabin probabilistic algorithm to test compositeness. It then gives a rigorous and self-contained treatment of Shor's algorithm. It also gives self-contained treatments of the algorithms of Grover and Simon.

Shannon made an extraordinarily deep and versatile foundation of information theory. The foundation is so solid that quantum information theory, despite its counterintuitive aspects, is a straightforward adaptation of Shannon's theory. Both Shannon's theory and its adaptation to the quantum setting are presented in detail.

There are deep and beautiful connections between quantum information theory and representation theory. I present this material, which is generally not available in book form. It includes an exposition of the quantum marginal problem, representation-theoretic proofs of standard entropic inequalities, and even a representation-theoretic proof of strong subadditivity.

0.1. "Real world" landscape as of 2024

The primary method that banks, governments, etc., currently use to communicate securely is the RSA cryptosystem.[1] RSA relies on the assumption that it is difficult to factor a large number N into its prime factors. In 1994 P. Shor [**Sho94**] (also see [**Sho97**]) described an algorithm to factor numbers quickly on a "quantum computer", something that at the time did not really exist. This and other developments spurred interest in building one.

Shortly after Shor's algorithm, Grover [**Gro96**] developed a remarkable search algorithm that is of complexity the square root of the size of a brute-force search (essentially the best-known classical algorithm for unstructured input). Despite a large stream of excellent research in the past 30 years, see, e.g., [**STF$^+$**], [**HHL09**], and [**AJSP14**], there have not been any "game-changing" discoveries comparable to Shor's or Grover's algorithms.

In January 2016, and in more detail in October, the NSA released a document warning the world that current encryption algorithms will no longer be secure as early as 2025[2]. They were concerned that by 2025 there might be quantum computers capable of effectively running Shor's algorithm. Such a development would endanger secure communication.

In response to Shor's algorithm, mathematicians, computer scientists, and cryptographers are developing *post-quantum cryptography*, a growing body of new cryptographic paradigms that are expected to be immune to attacks using a quantum computer.

Writing now, in 2024, there are working quantum computers. You may even test one yourself by going to `https://www.ibm.com/quantum/technology`.

On the other hand, it seems likely in the medium-term future that operating quantum computers will not be able to break RSA cryptography (but see [**KSB$^+$20**]). The stumbling block is that all existing large quantum computers are prone to errors, and Shor's algorithm is not fault tolerant.

Is there any near-term potential advantage of quantum computers over classical computers? The answer is yes, because the modeling of complex systems *is* fault tolerant and there is the hope that quantum computers may soon be better than classical computers in tasks such as designing and testing airplane wings. See, e.g., [**KEA$^+$23**]. The terminology in the literature for the current state is *noisy intermediate-scale quantum computing*, NISQ. Even if the advantage does not materialize, the challenges quantum computing presents

[1]In addition to RSA, the elliptic curve Diffie-Hellman method is also used, which relies on the assumption that computing the discrete logarithm is difficult. A variant of Shor's algorithm gives an efficient evaluation of the discrete logarithm. See [**Hen21**] for a discussion.

[2]See `https://people.tamu.edu/~jml//CNSA-Suite-and-Quantum-Computing-FAQ.pdf`. For the current warning, see `https://media.defense.gov/2022/Sep/07/2003071836/-1/-1/0/CSI_CNSA_2.0_FAQ_.PDF`.

to classical computing have spurred considerable research. One such example is that of *tensor networks*, which provide a method of classically simulating quantum processes. See, e.g., [**NLD**$^+$**23**] and the references therein.

Quantum information theory has brought new perspectives to several areas of research. For example, there have been recent advances in understanding the classical complexity of multiplying matrices by introducing ideas from quantum information theory [**CVZ19**, **CVZ18**].

0.2. Usage

For a one-semester graduate class, one can cover most of the book. For a one-semester undergraduate class, one can either cover Chapters 1–3 or cover up to and including §6.1 (the quantum noiseless channel theorem) while skipping the details of the algorithms of Miller-Rabin and Shor. It is also possible to give a short graduate course on either Chapters 1–3 on algorithms or Chapters 4–7 on information theory, using some material from Chapter 2.

Prerequisites. A background in undergraduate mathematics including some knowledge of elementary probability (probability distributions on finite sets), linear algebra (eigenvalues and eigenvectors of linear maps), and a semester of undergraduate algebra (basic definitions and properties of groups and rings) is needed.

Layout. All theorems, propositions, remarks, examples, etc., are numbered together within each section. For example, Definition 1.3.2 is the second numbered item in §1.3. Equations are numbered sequentially within each section. Hints and answers for exercises marked with the symbol ⊚ are given at the end of the book.

0.3. Acknowledgments

I am especially grateful to Alessandra Bernardi, who invited me to give a course on the subject at the University of Trento in the summer of 2017. I thank the Trento group for their wonderful hospitality, especially A. Bernardi, I. Carusotto, and L. Sola Conde. I thank M. Michałek for numerous suggestions after he taught a class at U. Konstanz from a draft of this book. I thank students C. Chang, R. Geng, P. Speegle, T. Tran, L. Wang, and D. Wu for extensive comments, corrections, and suggestions. I thank M. Christandl, I. Nechita, R. O'Donnell, and M. Walter for helping me learn the subject, and I thank S. Aaronson for advice on designing the course. I thank K. Dykema, M. Fanizza, I. Leigh, and especially G. Berkolaiko, P. Kuchment, J. M. Rojas, and my editor, S. Gelfand, for excellent comments on a near final version of this book.

0.4. Notation

What follows is notation that will be used throughout the book.

$i = \sqrt{-1}$.

$\omega_N = e^{\frac{2\pi i}{N}}$.

The natural logarithm is denoted ln and log denotes $\log_2$.

Binomial coefficients are $\binom{n}{p} := \frac{n!}{p!(n-p)!}$, and multinomial coefficients are $\binom{n}{p_1,\ldots,p_d} := \frac{n!}{p_1!\cdots p_d!}$.

For a complex number $z = x + iy$ or matrix X, $\bar{z} = x - iy$ and $\bar{X}$ denote their complex conjugates.

For functions f, g of a real variable (or integer) x:

$f(x) = O(g(x))$ if there exists a constant $C > 0$ and x_0 such that $|f(x)| \leq C|g(x)|$ for all $x \geq x_0$.

$f(x) = o(g(x))$ if $\lim_{x\to\infty} \frac{|f(x)|}{|g(x)|} = 0$.

$\mathrm{poly}(n)$ denotes a function whose growth is bounded above and below by polynomials in n.

$f(x) = \Omega(g(x))$ if there exists a constant $C > 0$ and x_0 such that $C|f(x)| \geq |g(x)|$ for all $x \geq x_0$.

Finite-dimensional complex vector spaces will generally be denoted U, V, W.

Finite-dimensional Hilbert spaces will generally be denoted $\mathcal{H}$, $\mathcal{H}_A$, $\mathcal{H}_B$, $\mathcal{H}_C$.

Subspaces of a Hilbert space will generally be denoted $\mathcal{M} \subset \mathcal{H}$.

For vector spaces V and W, V^* denotes the space of linear maps $V \to \mathbb{C}$, $\mathrm{Hom}(V, W) \cong W \otimes V^*$ denotes the space of linear maps from V to W, and $\mathrm{End}(V)$ denotes the space of linear maps from V to itself.

Elements of a Hilbert space $\mathcal{H}$ will generally be denoted $|x\rangle$ and elements of $\mathcal{H}^*$ will be denoted $\langle x|$. For $|v\rangle, |w\rangle \in \mathcal{H}$, $\langle v|w\rangle$ denotes their Hermitian inner product, which we take to be linear in the second factor and conjugate linear in the first. We use the notations $|xy\rangle := |x\rangle \otimes |y\rangle$ and $\langle xy| := \langle x| \otimes \langle y|$. The pairing $\mathcal{H} \otimes \mathcal{H}^* \to \mathbb{C}$ is written $|x\rangle \otimes \langle y| \mapsto \langle y|x\rangle$.

For a subspace $U \subset V$,

$$U^{\perp} := \{\alpha \in V^* \mid \alpha(u) = 0 \;\forall u \in U\}$$

denotes its annihilator in the dual space.

For a subspace $\mathcal{M} \subset \mathcal{H}$,

$$\mathcal{M}^{\perp} := \{|v\rangle \in \mathcal{H} \mid \langle w|v\rangle = 0 \;\forall |w\rangle \in \mathcal{M}\}$$

denotes its orthogonal complement.

$\mathcal{X} = \{a_1, \ldots, a_n\}$ denotes a finite set. We sometimes write $[n] = \{1, ..., n\}$.

When $\mathcal{X} = \{a_1, ..., a_n\}$ we let $p_{\mathcal{X}}$ denote a probability distribution on $\mathcal{X}$.

$\bar{p}_{\mathcal{X}} = (p_{\mathcal{X}}(a_1), \ldots, p_{\mathcal{X}}(a_n))$ denotes the vector representing the distribution $p_{\mathcal{X}}$.

$\Pr(M)$ is the probability that the event M occurs with respect to some understood probability distribution.

For a matrix $X \in \mathrm{Mat}_{n\times m}$, $X^{\mathbf{t}} \in \mathrm{Mat}_{m\times n}$ denotes its transpose and for a linear map $f : V \to W$, its transpose is denoted $f^{\mathbf{t}} : W^* \to V^*$, which is the induced linear map on dual spaces.

For Hilbert spaces $\mathcal{H}_A$ and $\mathcal{H}_B$ and an element $X \in \mathrm{Hom}(\mathcal{H}_A, \mathcal{H}_B)$, $X^\dagger \in \mathrm{Hom}(\mathcal{H}_B, \mathcal{H}_A)$ denotes the adjoint linear map defined by $\forall v \in \mathcal{H}_A, w \in \mathcal{H}_B$, $\langle Xv|w\rangle_{\mathcal{H}_B} = \langle v|X^\dagger w\rangle_{\mathcal{H}_A}$. In particular, for $X \in \mathrm{Mat}_{n\times n}(\mathbb{C})$, $X^\dagger := \bar{X}^{\mathbf{t}}$.

$\mathrm{Herm}(\mathcal{H}) \subset \mathrm{End}(\mathcal{H})$ is the subset of Hermitian endomorphisms; namely $X \in \mathrm{End}(\mathcal{H})$ such that $X^\dagger = X$.

$\mathcal{D}(\mathcal{H}) \subset \mathrm{Herm}(\mathcal{H})$ is the subset of density operators; see Definition 5.1.11.

$\mathbf{U}(n) := \{X \in \mathrm{Mat}_{n\times n}(\mathbb{C}) \mid X^\dagger = X^{-1}\}$ is the unitary group; see Definition 2.1.4.

$\mathbf{O}(n) := \{X \in \mathrm{Mat}_{n\times n}(\mathbb{R}) \mid X^{\mathbf{t}} = X^{-1}\}$ is the orthogonal group; see equation (2.1.3).

$\mathbb{C}^* = \mathbb{C}\backslash\{0\}$ is the nonzero complex numbers.

$\mathbb{F}_2$ is the field with two elements 0 and 1.

$\mathbb{Z}/M\mathbb{Z}$ denotes the ring of integers modulo M; see §A.1.

$a \bmod M$ denotes the equivalence class of a in $\mathbb{Z}/M\mathbb{Z}$ but we slightly abuse notation and use the same notation for the nonnegative integer that is the remainder when a is divided by M, i.e., the smallest nonnegative representative of the equivalence class.

For $Z \in \mathrm{End}(\mathcal{H})$ and $X \in \mathrm{End}(\mathcal{H})$, $Z \cdot X := ZXZ^\dagger$.

Chapter 1

Classical and probabilistic computation

This chapter presents classical and probabilistic computation via linear algebra in order to pave the way for quantum computation. It begins, in §1.1, with a vague presentation of several computational tasks to preview things to come. Next, in §1.2, two surprising algorithms for the simple task of multiplying integers are described: Napier's book of logs, which dramatically advanced astronomy, and the Schönhage-Strassen algorithm for multiplying polynomials and integers. The Schönhage-Strassen algorithm utilizes the fast Fourier transform (FFT) which will be essential for Shor's quantum factorizing algorithm. In §1.3, we discuss *circuits*, the basic classical model of computation. We take two steps towards quantum computation: we first introduce reversible classical computation in §1.4, because quantum computation must be reversible. Quantum algorithms will be expressed as a sequence of matrix-vector multiplications, so in §1.5, we rephrase classical computation in the language of linear algebra to facilitate comparisons. In §1.6, we discuss probabilistic computation phrased in the language of linear algebra. Finally, in §1.7, we discuss several relevant complexity classes and what is known about them.

1.1. Complexity of various tasks

Consider the following tasks and ask yourself how difficult they are to perform:

(1) Multiply two integers.

(2) Compute the determinant of a matrix with integer entries.

(3) Compute the permanent of a matrix with integer entries. (The permanent is the same polynomial as the determinant except that there are no minus signs.)

(4) Given a polynomial in several variables expressed as a circuit (see §1.3), determine if it is the zero polynomial.

(5) Determine if a given integer is prime.

(6) Factor an integer into prime factors.

(7) Simulate a complex system, e.g., weather or a chemical reaction.

How much work is involved in multiplying two integers depends on their size, so we are really asking how the difficulty grows as a function of their size. If we work in base 10 and count addition and multiplication of two single-digit numbers each as unit cost, then the naïve algorithm to compute mn uses $(1+\lceil\log_{10}(n)\rceil)(1+\lceil\log_{10}(m)\rceil)$ single-digit multiplications and a smaller number of additions. Usually in computer science one works in base 2 and counts addition and multiplication in $\mathbb{F}_2$ (i.e., single-digit operations in base 2) as unit cost, so the size of n is the number of digits in base 2, i.e., $\log(n)$, and the naïve algorithm has cost $O(\log(n)\log(m))$. Here and in general, $O(f(n))$ denotes a function bounded above by a constant times some given function $f(n)$ and the same notation is used for functions of several variables. In §1.2 we will see that already for this first task there are several surprises.

The standard formula for computing the determinant of an $n \times n$-matrix has $n!$ terms, so the work involved appears to grow exponentially with the size of the matrix. However, using Gaussian elimination (which amounts to a sequence of matrix multiplications), one can perform the task in $O(n^3)$ arithmetic operations. (One can do even better; see §1.2.4.) The formula for the permanent has the same number of terms, but no significant simplification of its computation is known, and it is conjectured **[Val79]** that there is none. This is a variant, due to L. Valiant, of the millennium prize problem P v. NP.

There is no known polynomial time deterministic algorithm for task (4). However, if one has access to randomness, one can simply evaluate the polynomial at a random point. If one obtains something nonzero, one knows the polynomial is not the zero polynomial. If one obtains zero, then with high probability (probability as close to one as desired if working over an infinite field), one concludes it is the zero polynomial. For more on this problem, see §1.7.2. Thus, at least for some problems, access to randomness may increase computing power. We say “may”, as it has not been proven that there is no polynomial time deterministic algorithm. The situation with quantum computation is similar. It is believed that access to a quantum computer would increase computing power, but so far no increase in computing power has been proven.

However, it has been proven that access to *entanglement* (a quantum resource discussed in §2.6 and throughout this book) increases communication power.

Testing if a number is prime at first glance does not seem so different from finding its prime factors. We will see that the state of the art for the two questions is quite different. We have already mentioned Shor's algorithm. We describe the situation for primality in §3.4, where we discuss a famous fast algorithm that utilizes access to randomness, and the current state of the art. Here there are several surprises.

The last problem is not clearly posed, but as mentioned in the preface, it holds the best medium-term promise for quantum computers performing better than classical computers.

1.2. Surprising algorithms

1.2.1. Logarithms: Fast multiplication of numbers. Until the 1600s, whenever people attempted to make astronomical predictions[1], a difficult step was the multiplication of large numbers. In 1614 John Napier revolutionized computation by writing a book to implement a transform that swaps multiplication for addition: the logarithm. Namely the book simply lists integers and numerical approximations of their logarithms. To multiply two large integers, one looks up their logarithms, adds them, and then looks up the sum in the table of logs and records its exponentiation. (If one wants the exact answer, one must take a sufficiently good approximation that the integer closest to the exponentiation is guaranteed to be the correct one.) Johannes Kepler used Napier's book to make astronomical tables on the order of 30 times more accurate than previous tables [**Gle11**, p. 87].

Even in this old example, there is something modern to learn: if the difficult step of a calculation can be precomputed and stored in a database, it becomes essentially "free".

Remark 1.2.1. Exact results may be obtained at the end by rounding to the nearest integer, even though the actual output is approximate. Quantum computing will also have this feature of being approximate and if one wants an exact answer, one must take a sufficiently good approximation so that one obtains an exact result by rounding.

Exercise 1.2.2. Pretend you have Napier's book in hand and you want to multiply two four-digit numbers, say 8,764 and 2,366. Use your computer to compute an approximation of their logs in base 10, add the approximate logs, and then raise 10 to that power. How accurate an approximation do you need to get the first four digits of the answer correct? To get the full correct answer?

[1]The king was very interested in knowing auspicious dates for action; see [**Lyo09**].

1.2.2. The DFT: Fast multiplication of polynomials. The famous mathematician Kolmogorov conjectured that one could not multiply two n-digit numbers using less than n^2 one-digit multiplications. His student Karatsuba showed in fact that one can use $O(n^{1.6})$ one-digit multiplications. We now describe an even more efficient algorithm for multiplying numbers due to Schönhage and Strassen [**SS71**] that also may be used to multiply polynomials. We first present the algorithm for polynomials. A variant of this algorithm will be critical to Shor's quantum algorithm for factoring.

Let $a(x), b(x)$ be polynomials of degree at most d. Write $a(x) = \sum_{i=0}^{d} a_i x^i$, $b(x) = \sum_{j=0}^{d} b_j x^j$. Write $\bar{a} = (a_0, \dots, a_d)^{\mathbf{t}}$ and similarly for $\bar{b}$, where here and throughout, $()^{\mathbf{t}}$ denotes transpose. Writing $a(x)b(x) = \sum_{k=0}^{2d} c_k x^k$, one has

$$c_k = \sum_{i+j=k} a_i b_j. \tag{1.2.1}$$

One says $\bar{c}$ is the *convolution* of $\bar{a}$ and $\bar{b}$. To obtain the coefficient vector $\bar{c}$ by this standard method, one needs to perform on the order of d^2 arithmetic operations (i.e., $+$'s and $*$'s, where $*$ denotes multiplication). Here we treat each scalar operation as unit cost and will do so for the Schönhage-Strassen algorithm. If the cost of integer multiplication depends on the size of the integer and the sizes of the integers involved are uniformly bounded, then in both algorithms the cost will be multiplied by a constant so the ratios of the costs will be unchanged.

To express this calculation in terms of matrix-vector multiplication, note that the vector $\bar{c}$ is the product of the following $(2d+1) \times (d+1)$-matrix and column vector of height $d+1$:

$$\begin{pmatrix} a_0 & 0 & & \cdots & 0 \\ a_1 & a_0 & 0 & \cdots & 0 \\ a_2 & a_1 & a_0 & \ddots & \\ \vdots & & \vdots & & \\ a_d & a_{d-1} & & \cdots & a_0 \\ 0 & a_d & a_{d-1} & \cdots & a_1 \\ \vdots & & \ddots & & \\ 0 & \cdots & & 0 & a_d \end{pmatrix} \begin{pmatrix} b_0 \\ b_1 \\ \vdots \\ b_d \end{pmatrix}.$$

Here we have broken the symmetry between $a(x)$ and $b(x)$. The symmetry will be restored momentarily.

In our first algorithm the input was transformed in such a way that the computation became easier and then transformed back. Here we will do something similar. We will perform a transformation that re-organizes the input data of the two polynomials to simplify the calculation. A difference will be that instead of being free, the cost of the transform will be the bulk of the cost of the algorithm.

Up until now, we have not mentioned what the coefficients of the polynomials are: they could be integers, rational numbers, real numbers, or complex numbers. To prepare for what comes next, we will allow complex numbers. Since $\deg(ab) \leq 2d$, instead of working in the space of all polynomials in one variable, denoted $\mathbb{C}[x]$, we can work in a *quotient* of this space, namely the ring $\mathbb{C}[x]/(x^N - 1)$ of polynomials quotiented by the ideal generated by the polynomial $x^N - 1$ for any $N > 2d$. See §A.1 for basic definitions regarding rings and groups and a brief discussion of $\mathbb{C}[x]/(x^N - 1)$. Elements of this ring are equivalence classes of polynomials, where $x^N \equiv 1$ generates the equivalence relation. For the moment, to fix ideas set $N = 2d + 1$, but later we will take N to be a power of two. We can then write a $(2d + 1) \times (2d + 1)$-matrix for $a(x)$ (allowing it now to have degree $2d$) as the matrix

$$\begin{pmatrix} a_0 & a_{2d} & a_{2d-1} & \cdots & a_3 & a_2 & a_1 \\ a_1 & a_0 & a_{2d} & \cdots & a_4 & a_3 & a_2 \\ a_2 & a_1 & a_0 & \ddots & & \vdots & \\ \vdots & & \vdots & & & & \\ a_d & a_{d-1} & & \cdots & a_{d+3} & a_{d+2} & a_{d+1} \\ a_{d+1} & a_d & a_{d-1} & \cdots & a_{d+4} & a_{d+3} & a_{d+2} \\ \vdots & & \ddots & & & & \\ a_{2d} & \cdots & & & a_2 & a_1 & a_0 \end{pmatrix} \tag{1.2.2}$$

and similarly for $b(x)$, although we only need the first column of the product. A matrix whose entries are cyclic permutations of the entries in the first column as in (1.2.2) is called a *circulant* matrix.

If we set $a_{d+1} = \cdots = a_{2d} = 0$, the first $d + 1$ columns of this matrix are our old matrix. This looks like we are making our problem more complicated. However, now that we have a square matrix, call it A, we can attempt to *diagonalize* it. The attempt will succeed; see Exercise 1.2.6 below. Call the diagonalized matrix $\hat{A}$. Even better, and to restore symmetry, we will write the corresponding matrix for $b(x)$ and see that A and B are simultaneously diagonalizable.

Notation. Throughout this book $i = \sqrt{-1}$, unless it is being used as an index. Write $\omega_N = e^{\frac{2\pi i}{N}}$.

Definition 1.2.3. The (unnormalized) *discrete Fourier transform* is the map $\mathrm{DFT}_N : \mathbb{C}^N \to \mathbb{C}^N$ given by the matrix whose (j, k)-entry is

$$(\mathrm{DFT}_N)_{jk} := \omega_N^{jk}. \tag{1.2.3}$$

The name *discrete Fourier transform* will be explained in Remark 1.2.13.

Remark 1.2.4. For reasons explained in Chapter 2, often it is better to deal with the normalized matrix with entries $\frac{1}{\sqrt{N}}\omega_N^{jk}$, as then the associated map is unitary.

The matrix DFT_N is an example of a *Vandermonde matrix*; see Definition 3.3.9.

The next proposition shows that DFT_N sends the coefficient vector of a polynomial of degree at most $N-1$ to a vector with entries that are the eigenvalues of the corresponding $N \times N$ circulant matrix (1.2.2).

Proposition 1.2.5. *Any $N \times N$ circulant matrix is diagonalizable by conjugating the matrix by the matrix* DFT_N *of* (1.2.3). *In particular, all $N \times N$ circulant matrices may be simultaneously diagonalized.*

We present two proofs:

Pedestrian proof.

Exercise 1.2.6. Show that for all $0 \leq j \leq N-1$, the vector

$$(1, \omega_N^j, \omega_N^{2j}, \dots, \omega_N^{(N-1)j})^{\mathbf{t}}$$

is an eigenvector of (1.2.2).

The entries of the eigenvector are independent of the a_s, so all circulant matrices have the same eigenvectors.

Finally recall that an $N \times N$-matrix A is diagonalizable if and only if it has N linearly independent eigenvectors, and a change of basis matrix is given by an $N \times N$-matrix whose columns are linearly independent eigenvectors of A. □

Geometric proof. First observe that $a(x)b(x) = b(x)a(x)$, in both the usual multiplication of polynomials and as elements of $\mathbb{C}[x]/(x^N - 1)$. The map

$$\mathbb{C}[x]/(x^N - 1) \to \mathrm{Mat}_{N \times N},$$

where $a(x)$ is sent to the matrix (1.2.2) and $\mathrm{Mat}_{N\times N}$ denotes the ring of $N \times N$-matrices (endomorphisms of $\mathbb{C}^N$), is a ring homomorphism, which ensures the corresponding matrices will commute. The image is exactly the space of circulant matrices. Since the algebra $\mathbb{C}[x]/(x^N - 1)$ is generated by x, it will be sufficient to prove the result for the matrix corresponding to multiplication by x. Thus it will be sufficient to show that for each k, the k-th column vector of (1.2.3) is an eigenvector for multiplication by x in the ring $\mathbb{C}[x]/(x^N - 1)$ with eigenvalue ω_N^{-k}. To see this consider (the equivalence class of)

$$x(1 + \omega_N^j x + \omega_N^{2j} x^2 + \cdots + \omega_N^{(N-1)j} x^{N-1})$$

and notice that $\omega_N^{(N-1)j}(x^N) \equiv \omega_N^{-j}(1)$. □

Given $\hat{A}, \hat{B}$, to compute the matrix $\hat{C} := \hat{A}\hat{B}$, we only need to perform $N = O(d)$ scalar multiplications instead of $O(d^3)$.

Exercise 1.2.7. Verify that if A, B are diagonalized by P, then $C = P^{-1}\hat{C}P$ is indeed the matrix corresponding to $\bar{c}$.

Exercise 1.2.8. Show that if two diagonalizable matrices commute, then they are simultaneously diagonalizable. ⊚

Exercise 1.2.9. Is the same true if one has three diagonalizable commuting matrices?

However, diagonalizing seems like a bad idea: the cost of a change of basis is worse than $O(d^2)$. Moreover, we will have to un-diagonalize $\hat{C}$ to get the coefficients of $c(x)$ (which are the elements of the first column of C). The punch line will be that for the highly structured matrices in our situation there exists a relatively cheap way to carry out the diagonalization.

Write $\hat{a} = \mathrm{DFT}_N \bar{a}$ (where we have padded the coefficient vector of $a(x)$ with zeros to make it have length N), and similarly $\hat{b} = \mathrm{DFT}_N \bar{b}$. Given $\hat{a}$ and $\hat{b}$, the vector $\hat{c}$ can be computed using N entrywise scalar multiplications as $\hat{c}_k = \hat{a}_k \hat{b}_k$. Finally $\bar{c} = \mathrm{DFT}_N{}^{-1}\hat{c}$.

Remark 1.2.10. For those familiar with the representation theory of finite groups, the discrete Fourier transform is the change of basis matrix from the standard basis of the regular functions on the cyclic group of order N to the character basis. The roots of unity appearing in the DFT matrix are the characters of the cyclic group. Using representation theory, it is straightforward to derive the DFT matrix.

Now we come to a great discovery of Gauss in 1810 [**Gau**], which was rediscovered by several people, including Danielson and Lanczos in 1942 [**DL42**] (the first published version of the discovery that we are aware of) and Cooley and Tukey in 1965 [**CT65**] (which led to its modern implementation that revolutionized signal processing): the DFT matrix factors as a product of sparse matrices. More precisely, if the matrices are $N \times N$ and N is a power of two, there will be $2\lceil \log(N) \rceil$ matrices, each of which has at most $2N$ nonzero entries. The factorization is called the *fast Fourier transform* (FFT).

To get intuition as to why the DFT matrix might admit such a factorization, assume N is even. The columns labeled by even values of j only have even powers of ω_N appearing and consist of a doubling of the matrix for $\mathrm{DFT}_{N/2}$ with the orders re-arranged. Similarly, the entries columns indexed by odd j are almost a doubling of the matrix for $\mathrm{DFT}_{N/2}$, except that in addition to the re-arrangement, the j-th column is also multiplied by ω_N^j. This is exactly what will be exploited.

Let $\{|j\rangle \mid 0 \leq j \leq N-1\}$ denote the standard basis of $\mathbb{C}^N$. Assume N is a power of 2. We have

$$\begin{aligned}
\mathrm{DFT}_N(|j\rangle) &= \sum_{k=0}^{N-1} \omega_N^{jk}|k\rangle \\
&= \sum_{k:\text{even}} \omega_N^{jk}|k\rangle + \sum_{k:\text{odd}} \omega_N^{jk}|k\rangle \\
&= \sum_{\ell=0}^{\frac{N}{2}-1} \omega_N^{2j\ell}|2\ell\rangle + \sum_{\ell=0}^{\frac{N}{2}-1} \omega_N^{j(2\ell+1)}|2\ell+1\rangle \\
&= \sum_{\ell=0}^{\frac{N}{2}-1} \omega_{N/2}^{j\ell}|2\ell\rangle + \omega_N^j \sum_{\ell=0}^{\frac{N}{2}-1} \omega_{N/2}^{j\ell}|2\ell+1\rangle.
\end{aligned}$$

We can thus compute DFT_N by twice applying the $\mathrm{DFT}_{N/2}$ matrix and multiplying by several sparse matrices. Let Δ_N be the diagonal matrix with entries $(1, \omega_N, \omega_N^2, \ldots, \omega_N^{\frac{N}{2}-1})$, so $-\Delta_N$ is the diagonal matrix with entries $(\omega_N^{\frac{N}{2}}, \ldots, \omega_N^{N-1})$. Let Π_N denote the permutation matrix corresponding to the permutation $(0, 1, \ldots, N-1) \mapsto (0, 2, 4, \ldots, N-2, 1, 3, 5, \ldots, N-1)$. We have the matrix equation

$$\mathrm{DFT}_N = \begin{pmatrix} \mathrm{Id}_{N/2} & \Delta_N \\ \mathrm{Id}_{N/2} & -\Delta_N \end{pmatrix} \begin{pmatrix} \mathrm{DFT}_{N/2} & \\ & \mathrm{DFT}_{N/2} \end{pmatrix} \Pi_N. \tag{1.2.4}$$

For example,

$$\mathrm{DFT}_4 = \begin{pmatrix} 1 & 1 & 1 & 1 \\ 1 & \omega_4 & \omega_4^2 & \omega_4^3 \\ 1 & \omega_4^2 & \omega_4^4 & \omega_4^6 \\ 1 & \omega_4^3 & \omega_4^6 & \omega_4^9 \end{pmatrix} = \begin{pmatrix} 1 & 1 & 1 & 1 \\ 1 & \omega_4 & \omega_2 & -\omega_4 \\ 1 & \omega_2 & 1 & \omega_2 \\ 1 & -\omega_4 & \omega_2 & \omega_4 \end{pmatrix},$$

while, when $N = 4$,

$$\begin{aligned}
&\begin{pmatrix} \mathrm{Id}_{N/2} & \Delta_N \\ \mathrm{Id}_{N/2} & -\Delta_N \end{pmatrix} \begin{pmatrix} \mathrm{DFT}_{N/2} & \\ & \mathrm{DFT}_{N/2} \end{pmatrix} \Pi_N \\
&= \begin{pmatrix} 1 & 0 & 1 & 0 \\ 0 & 1 & 0 & \omega_4 \\ 1 & 0 & -1 & 0 \\ 0 & 1 & 0 & -\omega_4 \end{pmatrix} \begin{pmatrix} 1 & 1 & 0 & 0 \\ 1 & \omega_2 & 0 & 0 \\ 0 & 0 & 1 & 1 \\ 0 & 0 & 1 & \omega_2 \end{pmatrix} \begin{pmatrix} 1 & 0 & 0 & 0 \\ 0 & 0 & 1 & 0 \\ 0 & 1 & 0 & 0 \\ 0 & 0 & 0 & 1 \end{pmatrix} \\
&= \begin{pmatrix} 1 & 1 & 1 & 1 \\ 1 & \omega_4 & -1 & -\omega_4 \\ 1 & -1 & 1 & -1 \\ 1 & -\omega_4 & -1 & \omega_4 \end{pmatrix},
\end{aligned}$$

where to get the last line and see the equality of the two expressions, recall that $\omega_2 = -1$.

Exercise 1.2.11. Verify (1.2.4) directly for DFT_6.

If we iterate (1.2.4), replacing the instances of $\mathrm{DFT}_{N/2}$ by three block matrices, then replacing the instances of $\mathrm{DFT}_{N/4}$ by three block matrices, etc., when $N = 2^k$ we obtain a product of k matrices, each with at most 2^{k+1} nonzero entries.

The cost of matrix-vector multiplication of an $N \times N$-matrix with at most ℓ nonzero entries in each row is $O(\ell N)$. Our sparse matrix has 2 nonzero entries in each row (and column), so the cost of performing our DFT is $O(N \log(N))$ instead of $O(N^2)$. The inverse of the DFT has the same cost by inverting the factorization.

We conclude:

Theorem 1.2.12 ([**SS71**], Schönhage-Strassen). *Assign multiplication of complex numbers unit cost. Then two polynomials of degree at most d may be multiplied with cost* $O(d \log(d))$.

Remark 1.2.13. You may have seen Fourier transforms of periodic functions, where convolution in the original space corresponds to multiplication in the transform space. This is the analogous transform when the group is the circle. Explicitly, write the unit circle in $\mathbb{R}^2$ as

$$S^1 = \{(\cos(\theta), \sin(\theta)) \mid \theta \in [0, 2\pi)\} \subset \mathbb{R}^2.$$

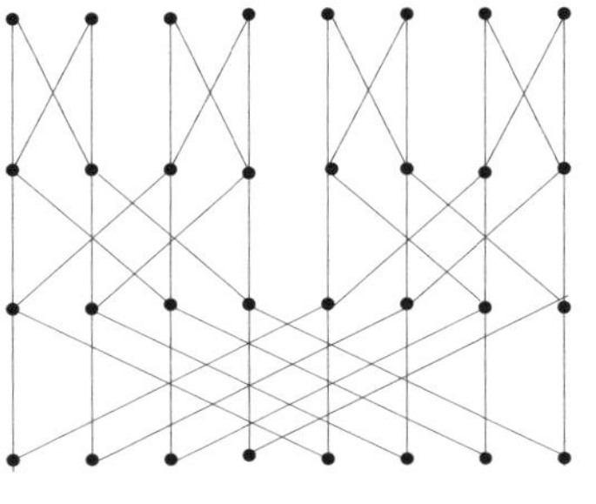

Figure 1.2.1. Sometimes the FFT is illustrated with this picture. The graph represents the product of three 8×8-matrices that gives DFT_8. One travels from the top to the bottom. Vertices in each row represent indices from 1 to 8, and edge from i to j at level $k \in \{1, 2, 3\}$ means the (i, j)-th entry of the k-th matrix is nonzero.

Introduce complex notation $\mathbb{C} = \mathbb{R}^2$, so $S^1 = \{e^{i\theta} \mid \theta \in [0, 2\pi)\} \subset \mathbb{C}$. Then for (e.g., continuous) functions $f(\theta)$ on the unit circle, we may write

$$f(\theta) = \sum_{n=-\infty}^{\infty} c_n e^{\frac{in\theta}{2}}, \quad \text{where } c_n = \frac{1}{2\pi}\int_0^{2\pi} f(\theta)e^{-\frac{in\theta}{2}}d\theta.$$

Since nonzero complex numbers form a group under multiplication and since the product of elements of length one is of length one, S^1 is naturally a group. In signal processing, one needs to digitize (e.g., sound waves), so one approximates a periodic function by sampling it at, say, N equally spaced points on the circle, e.g., the points $e^{\frac{k2\pi i}{N}}$, $0 \leq k \leq N-1$. These points form a subgroup, the cyclic group of order N. Tracing through the calculation, the DFT is the discretization of the Fourier transform on the circle, exactly what one needs in signal processing.

Aside 1.2.14. For those familiar with tensors and their ranks, the structure tensor of the algebra $\mathcal{A} = \mathbb{C}[x]/(x^N - 1)$, which may be written

$$T_{\mathcal{A}} = \sum_{k,j=0}^{N-1} x^k \otimes x^j \otimes e_{k+j \bmod N},$$

where e_j is the dual basis vector to x^j, has minimal tensor rank N, and the DFT is a change of basis that rewrites the structure tensor $T_{\mathcal{A}} \in \mathcal{A}^* \otimes \mathcal{A}^* \otimes \mathcal{A}$ as a sum of N rank one tensors.

Aside 1.2.15. One might wonder if there is an even more efficient way of computing the operation $\bar{a} \mapsto \mathrm{DFT}_N \bar{a}$. This question was posed by L. Valiant in [**Val77**], where he also gave a path to resolving it (he conjectured that no significant improvement is possible). There is interesting algebraic geometry related to the question; see [**KLPSMN09, GHIL16**].

1.2.3. Fast integer multiplication.

Assign multiplication and addition of single-digit integers unit cost. (We can work base in 10 or base 2; the comparison between algorithms will be the same.) If we want to multiply $m * n$ which, respectively, have d_m, d_n digits, in the naïve algorithm we need to perform $d_m d_n$ single-digit multiplications and a smaller number of additions, so the complexity of multiplying two d-digit numbers is $O(d^2)$.

If we apply the Schönhage-Strassen algorithm with x replaced by 10 if we work in base 10 or by 2 if we work in base 2, at first it looks like we can get away with $O(d\log(d))$ operations, but in intermediate steps one needs to work with complex numbers $a + bi$ where a, b are not necessarily integers. One fix is to use the FFT instead on a cyclic group of prime order for a prime of the form $2^m + 1$. Another is to represent the complex numbers by sufficiently good approximations. The actual cost using the algorithm is $O(d\log(d)\log(\log(d)))$.

Schönhage-Strassen conjectured that $O(d\log(d))$ cost was possible, and over 50 years later this was proved in [**HvdH21**]. This $O(d\log(d))$ cost was conjectured to be optimal and this question is still open.

For an in-depth discussion on the use of the FFT for integer and polynomial multiplication, see [**vzGG13**].

1.2.4. Matrix multiplication. Another surprising algorithm deals with matrix multiplication. The usual algorithm for multiplying two $n \times n$-matrices uses $O(n^3)$ arithmetic operations. Strassen [**Str69**] discovered an algorithm that uses $O(n^{2.81})$ arithmetic operations, and it has been conjectured that as n grows, it becomes nearly as easy to multiply matrices as it is to add them. That is, for any $\epsilon > 0$, one can multiply matrices using $O(n^{2+\epsilon})$ arithmetic operations. By 1988 it had been shown that $O(n^{2.38})$ is possible [**CW90**], but since then progress has been limited to an improvement by 0.004 [**Sto, Wil, LG14, AW21, VXXZ23**].

1.3. Classical computation via circuits

Classical complexity works in binary: one deals with strings of 0's and 1's. An element of the set $\{0,1\}$ is called a *bit*: it can encode "one bit" of information. It requires $1 + \lceil \log(N) \rceil$ bits to express a positive integer N in binary.

Definition 1.3.1. A *Boolean function* is a map $f : \mathbb{F}_2^n \to \mathbb{F}_2$, or more generally a map $\mathbb{F}_2^n \to \mathbb{F}_2^m$.

We agree on some basic Boolean functions whose complexity is designated as having unit cost, e.g.,

- addition $\oplus$ (also called *XOR*) where $a \oplus b$ is addition in $\mathbb{F}_2$, i.e., $0 \oplus 0 = 1 \oplus 1 = 0$ and $0 \oplus 1 = 1 \oplus 0 = 1$,
- $\neg$ (NOT) negation, which swaps 0 and 1,
- (OR) $a \vee b$ where $0 \vee 0 = 0$ and all other $a \vee b = 1$,
- (AND = multiplication in $\mathbb{F}_2$) $a \wedge b = ab$, where $1 \wedge 1 = 1$ and all other $a \wedge b = 0$.

Such a collection is called a *(logic) gate set*.

Definition 1.3.2. A *(Boolean) circuit* C is a finite, directed, acyclic graph with a collection of vertices of in-degree 0, a collection of vertices of out-degree 0, and other vertices that have positive in-degrees and out-degrees. The vertices of in-degree 0 are labeled by elements of $\mathbb{F} \cup \{x_1, ..., x_n\}$, where the x_i are variables, and are called *inputs*. Those of positive in-degree are labeled with Boolean functions from a gate set and are called *gates*. If the out-degree of v is 0, then v is called an *output gate*.

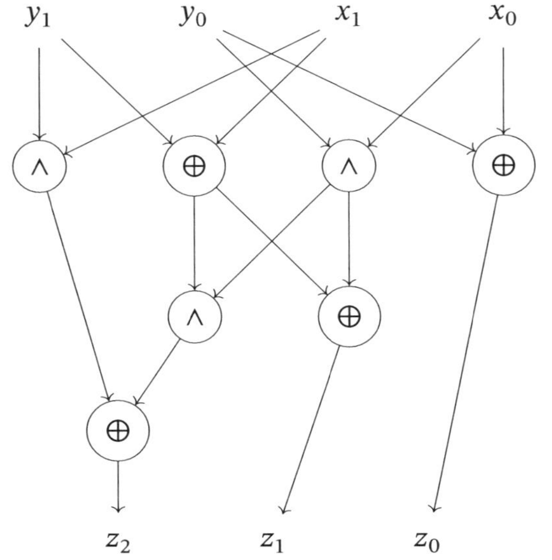

Figure 1.3.1. Circuit for $z = y + x$, where $y = y_1 2^1 + y_0 2^0$, etc. Output legs are the three digits of $z = z_2 2^2 + z_1 2^1 + z_0 2^0$.

Remark 1.3.3. See [**Fey01**, Chap. 2] for a discussion of how physical computers carry out Boolean operations.

Figure 1.3.1 depicts a Boolean circuit for the addition of two 2-digit (in binary) numbers.

Definition 1.3.4. A gate set is a *universal gate set* if any Boolean function can be computed with a circuit whose vertices are labeled with gate set elements.[2]

Proposition 1.3.5. *The gate set* $\{\neg, \wedge\}$ *is universal.*

Exercise 1.3.6. Give an explicit expression for $x \vee y$ and $x \oplus y$ in terms of the gate set $\{\neg, \wedge\}$.

Exercise 1.3.7. Prove Proposition 1.3.5 for Boolean functions that take the value 1 on exactly one input. ⊚

Exercise 1.3.8. Prove Proposition 1.3.5 in general. ⊚

Exercise 1.3.9. Write a Boolean circuit that computes $x * y$ when x, y are two digits in binary.

Exercise 1.3.10. What is the precise number of $\oplus$ and $\vee$ needed for a circuit to compute the product of integers M and N with the naïve algorithm?

[2]In some of the literature a gate set is sometimes called a "basis" (despite being unrelated to bases of vector spaces), and a universal gate set is called a "complete basis".

Using the naïve algorithm, we may compute M^k using a circuit of size $O(\log(M)^k)$.

Proposition 1.3.11. *If a, b are natural numbers with $a > b$, then the division of a by b and its remainder may be computed with a circuit of size $O(\log(a)^2)$.*

Proof. Express a and b in binary. Elementary long division requires one to compare the size of b and various numbers with $\log(b)$ digits, which costs $O(\log(b))$ each time. After the comparison, one may need to perform a subtraction, which also costs $O(\log(b))$, and there are at most $O(\log(a))$ steps. At the end, one obtains both $\lfloor \frac{a}{b} \rfloor$ and the remainder. □

Exercise 1.3.12. Say $M < N$ and instead of wanting to compute M^k, we are only interested in its remainder after division by N, i.e., to compute $M^k \bmod N$. Show that this can be computed with cost $O(k \log(N)^2)$ (without using Schönhage-Strassen). Moreover, if $k = 2^\ell$, then one can do the computation with cost $O(\ell \log(N)^2)$. ⊚

1.4. Reversible classical computation

We will see that the gates of a quantum circuit (other than the measurements) must be *reversible*; i.e., there must be another gate that takes the output of a gate and produces its input. In this section we show how classical computation may be made reversible.

In reversible computation, one deals exclusively with invertible maps $f : \mathbb{F}_2^n \to \mathbb{F}_2^n$, even if the result of interest is just the first entry of the output.

Of the gates we saw, NOT is clearly reversible as $\neg\neg x = x$. At the cost of using an extra bit, one can make addition and multiplication reversible as follows: consider the following gate, called the *Toffoli gate* $\text{Tof} : \mathbb{F}_2^3 \to \mathbb{F}_2^3$:

$$|x, y, z\rangle \mapsto |x, y, z \oplus (x \wedge y)\rangle. \tag{1.4.1}$$

If we send in $|x, y, 0\rangle$, we obtain $x \wedge y$ in the third entry, and if we send in $|x, 1, z\rangle$, we obtain $x \oplus z$. In the literature, the entries are often called *registers*, and we henceforth adopt this terminology.

Exercise 1.4.1. Show that $\text{Tof} \circ \text{Tof} = \text{Id}$, so Tof is indeed reversible.

Exercise 1.4.2. The *reversible controlled not* gate CNOT takes a pair x, y and if $x = 0$, it sends y to y, and if $x = 1$, it sends y to $\neg y$, and in both cases it sends x to itself. Write down CNOT as a reversible gate using our gate set and verify that it is reversible

The gate set $\{\text{Tof}, \neg\}$ is universal and reversible, so there is no loss in computing power if one restricts to reversible classical computation.

Exercise 1.4.3. Show that in fact $\{\text{Tof}\}$ alone is universal.

Implementing $\oplus$ or $\wedge$ with the Toffoli gate requires "workspace bits", which are bits that are not part of the input or output of the function but are needed to carry out the computation. Usually they are 0 at the beginning of a computation. To execute a Boolean function $\mathbb{F}_2^m \to \mathbb{F}_2^n$ in a Boolean circuit, one needs $\max\{m, n\}$ bits to carry it out, but with reversible computation, in general more bits are needed.

Remark 1.4.4. As early as 1961, researchers were concerned that as computers got more powerful, they would generate too much heat from erasing bits; see, e.g., [**Lan61**] and [**Ben03**]. One way to solve this problem would be to have reversible computation. Thus one could argue for reversible computation independent of quantum computation. See [**Fey01**, Chap. 5] for a discussion of how one may make the heat generated by a computation arbitrarily small using reversible classical computation.

1.5. Classical computation via linear algebra

To facilitate the comparison with quantum computing, we will encode classical computation as a sequence of matrix-vector multiplications (following [**AB09**, Exercise 10.4]). Readers not familiar with tensors should read section §A.2 before reading this section. Fix a field $\mathbb{F}$. For classical computation we could take $\mathbb{F} = \mathbb{F}_2$, but to facilitate the comparison with probabilistic computation we will take $\mathbb{F} = \mathbb{R}$. For quantum computation we will take $\mathbb{F} = \mathbb{C}$.

1.5.1. Induced bases in tensor products of spaces. Give $\mathbb{R}^2$ the basis $|0\rangle, |1\rangle$, which induces the basis $|i\rangle \otimes |j\rangle$, $i, j \in \{0, 1\}$ of $(\mathbb{R}^2)^{\otimes 2}$. Introduce the notation $I := (i_1, \dots, i_N)$ with $i_\alpha \in \{0, 1\}$. The basis of $\mathbb{R}^2$ also induces a basis of $(\mathbb{R}^2)^{\otimes N}$:

$$|I\rangle := |i_1\rangle \otimes \cdots \otimes |i_N\rangle, \quad i_\alpha \in \{0, 1\}, \quad 1 \leq \alpha \leq N.$$

If our bit string is 00101100, we represent it by the vector $|00101100\rangle \in \mathbb{R}^{2^8}$.

Now $(\mathbb{R}^2)^{\otimes N}$ is isomorphic as a vector space to $\mathbb{R}^{2^N}$, and sometimes it is also convenient to represent elements of $(\mathbb{R}^2)^{\otimes N}$ as column vectors, forgetting the tensor product structure. There is no canonical change of basis from the basis $|I\rangle$ to the standard basis of column vectors. We will use the lexicographic order to obtain a change of basis, so the vector

$$a|00\rangle + b|01\rangle + c|10\rangle + d|11\rangle \in \mathbb{R}^2 \otimes \mathbb{R}^2 \simeq \mathbb{R}^4$$

will be mapped to the column vector

$$\begin{pmatrix} a \\ b \\ c \\ d \end{pmatrix}.$$

The linear map $\begin{pmatrix} 0 & 1 \\ 1 & 0 \end{pmatrix} \otimes \mathrm{Id}_2$, which applies NOT on the first $\mathbb{R}^2$ and is the identity on the second $\mathbb{R}^2$, may be expressed as the 4×4-matrix

$$\begin{pmatrix} 0 & 0 & 1 & 0 \\ 0 & 0 & 0 & 1 \\ 1 & 0 & 0 & 0 \\ 0 & 1 & 0 & 0 \end{pmatrix}.$$

Exercise 1.5.1. Consider $\mathbb{C}^N \otimes \mathbb{C}^M$ with basis $|js\rangle$, $0 \leq j, k \leq N-1$, $0 \leq s, t \leq M-1$ and where each index pair (js) is viewed as an element of $G := \mathbb{Z}/N\mathbb{Z} \times \mathbb{Z}/M\mathbb{Z}$. If we encode the multiplication in G via matrix-vector multiplication or matrix-matrix multiplication as in §1.2.2, we get a complicated matrix. However, here we may also perform the DFT for G to get a diagonal matrix. What is the linear map DFT_G that executes this discrete Fourier transform? ⊚

Exercise 1.5.2. If $G = (\mathbb{Z}/2\mathbb{Z})^k$, what is the linear map/matrix $\mathrm{DFT}_G : (\mathbb{C}^2)^{\otimes k} \to (\mathbb{C}^2)^{\otimes k}$ that executes this transform? ⊚

1.5.2. Reversible classical computation via linear algebra. Let $f : \mathbb{F}_2^n \to \mathbb{F}_2^m$ be a Boolean function. We phrase the computation $x \mapsto f(x)$ as a sequence of restricted linear operations on a vector space $(\mathbb{R}^2)^{\otimes N}$, where $N = \max\{n, m\} + s$, where s is the number of workspace bits needed. The restrictions will be the following:

(1) Each matrix must be invertible and take vectors representing sequences of bits, i.e., basis vectors, to basis vectors. Such matrices in $\mathrm{End}(\mathbb{R}^{2^n})$ are the *permutation matrices*. That is, each row and column has exactly one entry equal to one and all other entries zero.

(2) In order to deal with finite gate sets, require that each matrix only alters a small number of entries. Since we have a universal gate set that uses at most three entries at a time, we will assume each matrix acts on at most some $(\mathbb{R}^2)^{\otimes 3}$ and is the identity on all other factors in the tensor product.

Each map will realize some Boolean gate. For example, say we want to realize the Toffoli gate

$$|x, y, z\rangle \mapsto |x, y, z \oplus (x \wedge y)\rangle.$$

In the lexicographical basis $|000\rangle, |001\rangle, |010\rangle, |011\rangle, |100\rangle, |101\rangle, |110\rangle, |111\rangle$, of $\mathbb{R}^8$, the matrix is

$$\begin{pmatrix} 1&0&0&0&0&0&0&0\\ 0&1&0&0&0&0&0&0\\ 0&0&1&0&0&0&0&0\\ 0&0&0&1&0&0&0&0\\ 0&0&0&0&1&0&0&0\\ 0&0&0&0&0&1&0&0\\ 0&0&0&0&0&0&0&1\\ 0&0&0&0&0&0&1&0 \end{pmatrix}. \tag{1.5.1}$$

Call this matrix the *Toffoli matrix.*

The negation gate $\neg$ is obtained with the linear map $\mathbb{R}^2 \to \mathbb{R}^2$ given by the matrix

$$\begin{pmatrix} 0&1\\ 1&0 \end{pmatrix}.$$

Exercise 1.5.3. Write the 8×8-matrices for $(x, y, z) \mapsto (x, y, z \oplus (x \oplus y))$ and $(x, y, z) \mapsto (x, y, z \oplus (x \vee y))$.

The computation of a Boolean function $f : \mathbb{F}_2^n \to \mathbb{F}_2^m$ via a reversible classical circuit has as its input the basis vector $|x, I\rangle \in \mathbb{R}^{2^{n+s}}$, where x is the input and I some initialization of the workspace bits that does not depend on x, usually just a string of zeros, and the output is of the form $|y, J\rangle$ where $y \in \mathbb{F}_2^m$ is the answer (and J will in general depend on x).

Since the gate set used is universal, any Boolean function may be computed by a reversible classical circuit. In other words, all permutations on the first n symbols may be obtained from products of Toffoli matrices. An important question discussed in §1.7 asks which functions may be computed *efficiently* with reversible classical computation.

1.6. Probabilistic computation via linear algebra

We now explain how to incorporate the additional resource of randomness in computation. We continue to use the language of linear algebra to express the computation.

Readers not familiar with basic definitions and notation from probability should now read §B.1.

Consider a set injection $\{0,1\}^m \to \mathbb{R}^{2^m}$, via the map $I \mapsto |I\rangle$, where I is a bit string of length m. A probability distribution on $\{0,1\}^m$ may be encoded as a vector in $\mathbb{R}^{2^m}$: give $\mathbb{R}^2$ basis $|0\rangle, |1\rangle$ and $(\mathbb{R}^2)^{\otimes m} = \mathbb{R}^{2^m}$ basis $|I\rangle$ where $I \in \{0,1\}^m$. Assign to the distribution p_I on $\{0,1\}^m$ (where $I \in \{0,1\}^m$) the vector $v = \sum_I p_I |I\rangle \in \mathbb{R}^{2^m}$.

To prepare for the analogy with quantum computation, define a *pbit*, i.e., a probabilistic bit, to be an element of the set

$$\{p_0|0\rangle + p_1|1\rangle \mid p_j \in [0,1] \text{ and } p_0 + p_1 = 1\} \subset \mathbb{R}^2.$$

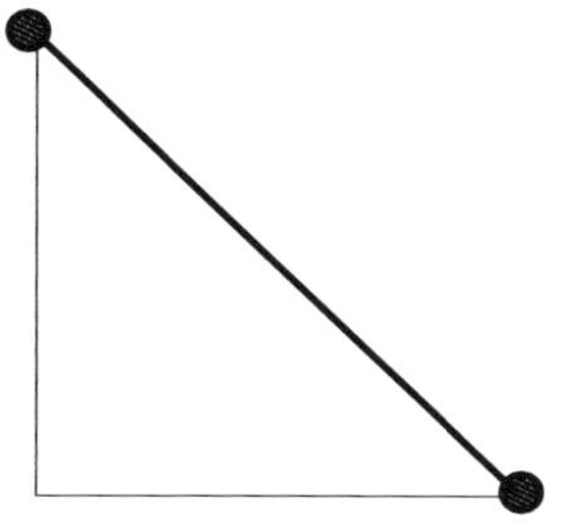

Figure 1.6.1. The set of pbits

In other words, the set of pbits is the set of probability distributions on $\{0,1\}$. The set of pbits is a convex set, and the basis vectors, i.e., the classical bits, are the extremal points of this convex set.

For probabilistic computation via linear algebra we require the following:

(1) Each linear map must take vectors representing probability distributions to vectors representing probability distributions. This implies the matrices are *stochastic*: the entries are nonnegative and the entries in each column sum to 1.

(2) In order to deal with finite gate sets, require that each linear map only alters a small number of entries. For simplicity assume it alters at most three entries; i.e., it acts on at most $(\mathbb{R}^2)^{\otimes 3}$ and is the identity on all other factors in the tensor product.

Exercise 1.6.1. Verify that the stochastic matrices are exactly those matrices that take probability distributions to probability distributions.

Since permutation matrices are stochastic, probabilistic computation includes classical computation.

To represent a fair coin flip in terms of linear algebra, introduce the matrix

$$\begin{pmatrix} \frac{1}{2} & \frac{1}{2} \\ \frac{1}{2} & \frac{1}{2} \end{pmatrix}.$$

Classical probabilistic computation cannot be made reversible, indicated by the fact that the coin flip matrix is not invertible.

For a probabilistic gate set one could take the coin flip and Toffoli matrices.

With this gate set, the matrices will actually be *doubly stochastic*: in addition to stochastic, the entries in each row will also sum to 1. The set of doubly stochastic matrices has a well-studied structure that we discuss in detail in §6.5.1.

Consider the case of a probabilistic algorithm where probability distributions with denominators that are not powers of two appear. When this probabilistic algorithm is expressed in our gate set, the computation will be *approximate*. An analogous phenomenon occurs for quantum computation.

Exercise 1.6.2. In probabilistic algorithms, we will want to choose an element uniformly at random from a set of M elements. How can we approximately realize this choice with our probabilistic gate set with error at most some fixed $\epsilon > 0$? (We can realize it exactly when M is a power of two.) ⊚

Thus with our probabilistic gate set, we can approximate any probability distribution arbitrarily closely by elements of the gate set and can perform all Boolean operations.

Summary of probabilistic computation via linear algebra: A probabilistic computation of $f : \mathbb{F}_2^n \to \mathbb{F}_2^m$, expressed in terms of linear algebra, starts with $|x0^{r+s}\rangle$ (or $|x0^{m-n+r+s}\rangle$ if $m > n$) where $x \in \mathbb{F}_2^n$ is the input. A sequence of elements from the gate set is applied to it, resulting in a vector $|v\rangle = \sum p_I |I\rangle$ that encodes a probability distribution on $\{0,1\}^{\max\{n,m\}+s+r}$. Then one takes the marginal distribution on the first m copies of $\mathbb{R}^2$: write $I = (i_1, \dots, i_{n+r+s}) = (J, L)$, where $J = (i_1, \dots, i_m)$, to obtain

$$\sum_J \sum_L p_{J,L} |J\rangle \in \mathbb{R}^{2^m}.$$

Here r is the number of coin flips needed and s is the number of workspace bits needed. The algorithm then outputs $J \in \{0,1\}^m$ with probability $\sum_L p_{J,L}$ using some external device that chooses J with probability $\sum_L p_{J,L}$.

In contrast with classical computation, we may not arrive at an arbitrary stochastic matrix via products of matrices corresponding to elements in our gate set. We may only approximate any such.

1.7. Brief overview of classical, probabilistic, and quantum complexity classes

1.7.1. P/poly, P, and NP.

We said that factorization is not known to have an efficient algorithm. We now make the notion of efficient precise.

Fix a universal gate set G. A natural complexity measure for a Boolean function is the minimal size of a Boolean circuit that computes it. Let $F_n : \mathbb{F}_2^n \to \mathbb{F}_2$ be a sequence of functions (F_n). Consider the growth with n of the minimal size of a circuit needed to compute F_n. If it grows like a polynomial, the sequence (F_n) is said to be in the class **P**/poly with respect to G. One says that sequences in **P**/poly have *efficient* algorithms.

Exercise 1.7.1. Show that membership in **P**/poly with respect to G is independent of the choice of the finite universal gate set G.

Thus one just says that the sequence (F_n) is in the class **P**/poly.

The complexity class **P** is the standard model for feasible computations, and it is unfortunate that **P**/poly, with a simple description, has a different definition than **P**. The definition of **P**/poly is less restrictive, but not so much less, and it can sometimes be used as a substitute for **P**; see [**AB09**, Chap. 6]. It is known that $\mathbf{P} \subsetneq \mathbf{P}/\text{poly}$; however, membership in **P**/poly for many problems of interest is not known.

The class **P** is usually defined in terms of a different model of computation, namely *Turing machines*. We will avoid defining them and assume the reader has at least a passing familiarity with them. A sequence (F_n) is in **P** if it is in **P**/poly and there exists a Turing Machine TM such that the circuits C_n computing F_n are constructed by TM in time poly(n); see [**KSV02**, Thm. 2.3]. Sometimes one takes the more restrictive definition of *efficient* that requires membership in **P**.

The complexity class **NP** essentially consists of problems whose proposed solutions can be verified quickly, i.e., in polynomial time. A famous problem in **NP** is the traveling salesman problem: if someone claims to have a route to visit 30 cities traveling less than 2,000 miles, it is easy to verify the claim by examining the route, but the only known way of *finding* such a route is essentially by a brute-force search. Another problem in **NP** is "SAT": one is handed a Boolean circuit and wants to know if it ever outputs 1. (If it does, to convince you it does, someone just needs to hand you an input that works, and you can quickly check if it outputs 1.) SAT and the traveling salesman problems are **NP**-*complete*, which means one could define **NP** to be the collection of problems that can be *reduced* (in polynomial time) to SAT; see, e.g., [**AB09**, Chap. 2]. In other words, there exists a polynomial time algorithm for SAT if and only if $\mathbf{P} = \mathbf{NP}$.

A problem like factoring asks for something beyond a $\{0, 1\}$ output and the corresponding complexity classes have slightly different names, but since we will not be discussing them further and since the corresponding issues are the same, we leave it to the curious reader to work through the zoo of complexity classes; see, e.g., `https://complexityzoo.net/Complexity_Zoo`.

1.7.2. BPP. It might increase our computational power if we exploit randomness. (Assuming we have a method to generate random numbers — more on this in Remark 4.5.14.) For example, if someone hands you a complicated expression for a polynomial, it can be very difficult to determine if the polynomial is just the zero polynomial in disguise. If we test the polynomial at a point and its evaluation is nonzero, then we know it is not the zero polynomial. If it does evaluate to zero, then we have no information. For a polynomial of degree d in one variable, it is sufficient to test $d + 1$ distinct points, but as the number of variables grows, the number of points one needs to check grows exponentially. However, if we are allowed to test at a random point and it evaluates to zero, then with high probability the polynomial is the zero polynomial. Even over finite fields, we can make this probability as high as we want by testing on several random points.

These observations motivate the class **BPP** (short for "bounded-error probabilistic polynomial time"), where one works with a Turing machine with access to randomness, and instead of asking for a correct yes or no answer on any input in polynomial time, one asks for a correct yes/no answer with probability strictly greater than $\frac{2}{3}$ on any input in polynomial time. (Thus if one runs the program enough times, one can get a correct answer on any input with probability as high as one wants.) See, e.g., [**KSV02**, §1.4] for the precise definition.

Remark 1.7.2. It is subtle to know if a given sequence of numbers is random. For example, the first digit of the number 2^n is far from a random element of $\{1,\dots,9\}$; see [**Ad89**, §16, Ex. 4]. Fortunately, for most situations, *pseudo-random* numbers suffice; see [**AB09**, §9.2.3].

Aside 1.7.3. If we are given certain types of additional information about a set of polynomials that we want to test, then one can test if a polynomial in the set is zero by testing a reasonable number of points. This subject, called PIT (polynomial identity testing), is an active area of research; see [**AB09**, §7.2.3]. For a geometric perspective see [**Lan17**, §7.7].

1.7.3. What is known. $\mathbf{P} \subseteq \mathbf{BPP} \subseteq \mathbf{P}/\mathrm{Poly} = \mathbf{BPP}/\mathrm{Poly}$.

The inclusion $\mathbf{BPP} \subseteq \mathbf{P}/\mathrm{Poly}$ is Adelman's theorem [**Adl78**]. The key observation is that "off-line" computations are not counted in the complexity assessment for **P**/Poly. So one can create, for any given n, a library of "random" a's to test on.

Does randomness really help? At the moment, we don't know. See [**AB09**, Chap. 20] for a discussion, where they state (p. 126) "most complexity theorists actually believe that **BPP** = **P**."

1.7.4. BQP.

We are not yet in a position to define the class **BQP**. It will be the quantum analog of **BPP**, the decision problems that can be solved efficiently, with high probability, on a quantum computer. Pbits will be replaced by *qubits*, which are unit vectors in $\mathbb{C}^2$ subject to an equivalence relation. The matrices will be allowed to have complex entries; they will be required to be *unitary* (see

§2.1) instead of stochastic. There is no generally used analog of **P** for a quantum computer as in general answers have a nonzero probability of being incorrect.

In a similar manner to how a finite probabilistic gate set can only approximate arbitrary distributions, a finite "universal quantum gate set" will only be able to approximate any unitary map arbitrarily closely by elements of the gate set.

1.7.5. The Church-Turing theses. The Church-Turing thesis (made explicitly by Church in [**Chu36**]) is:

Any algorithm can be realized by a Turing machine.

So far there has been no challenge to this thesis — e.g., any computation that can be done on a quantum computer can be done with a sufficiently large Turing machine. (Since we do not have a precise definition of what is allowable as an "algorithm", it would be impossible to affirmatively prove the thesis.)

The quantitative (sometimes called *strong*) Church-Turing thesis [**VSD86**] is:

Any algorithmic process can be simulated efficiently by a Turing machine

or

Any algorithmic process can be simulated efficiently by a probabilistic Turing machine.

If one believes factoring is not in **P**, then Shor's algorithm challenges this thesis. On the other hand, there are experts who think that factoring could be in **P**, because unlike, say, SAT or the traveling salesman problem, the problem is highly structured.

Chapter 2

Quantum mechanics for quantum computation

This chapter covers aspects of quantum mechanics needed for quantum computation.

It begins, in §2.1, with a list of ways to potentially improve probabilistic computation ("generalized probability") and first steps towards the potential improvements. This requires introducing Hilbert space and the unitary group. A brief discussion of the origins of quantum mechanics is given in §2.2. The postulates of quantum mechanics, which were arrived at via physical observations, are presented in §2.3 and are compared with the proposed "generalized probability". The chapter concludes with three interesting phenomena arising in quantum mechanics relevant for quantum information theory: super-dense coding in §2.4, "quantum teleportation" in §2.5, and the Einstein-Podolsky-Rosen "paradox" resolved by Bell in §2.6.

2.1. "Improved" probability and the linear algebra associated to it

This section discusses how one could potentially improve the computational power of probabilistic computation and make it more "physically natural" with the use of "better" gates. It is inspired by the presentation in [**Aar13**].

Machines that perform classical computation have been around for some time. Assuming "sufficiently random" rather than "truly random" computation is sufficient, probabilistic computers may also be built. In the wish list that follows we ignore whether a computer can be built carrying out generalized probabilistic computation.

2.1.1. A wish list for generalized probability. In §1.6 we saw that any Boolean function $f : \{0,1\}^n \to \{0,1\}^m$ that could be computed correctly with probability at least $\frac{2}{3}$ on any $I \in \{0,1\}^n$, using s workspace bits and r coin flips, could be computed with the same probability via a sequence of linear operators on $(\mathbb{R}^2)^{\otimes n+r+s}$. Each linear operator was stochastic (in fact doubly stochastic), taking vectors representing probability distributions to vectors representing probability distributions, and each operator acted on at most three registers. To get the output, after performing the linear operations, we took the marginal distribution to get a probability distribution on the first m entries of the output vector, and the algorithm outputs a basis vector in the first $(\mathbb{R}^2)^{\otimes m}$. The probability of I being the output is the coefficient of $|I\rangle$ in the marginal distribution.

Here is a wish list for how we might want to improve upon this set-up:

(1) Have more computing power by introducing more general types of linear maps as gates.

(2) Make probabilistic computation reversible (see the discussion in §1.4).

(3) Allow for a "continuous" evolution of the probability vector. More precisely, have the probability vector depend continuously on a parameter t such that if $|\psi_{t_0+t_1}\rangle = X|\psi_{t_0}\rangle$, where X is an admissible linear map, then there exist admissible linear maps Y, Z such that $|\psi_{t_0+\frac{1}{2}t_1}\rangle = Y|\psi_{t_0}\rangle$, $|\psi_{t_0+t_1}\rangle = Z|\psi_{t_0+\frac{1}{2}t_1}\rangle$, and $X = ZY$.

The motivation for the first wish is clear, and the second has already been discussed. To motivate the third, physical systems evolve with time and we would like to be able to approximate that time evolution. Moreover, having such a property will facilitate approximation as will be made precise in §3.5.

Let's start with wish (2). One way to make the coin flip reversible is that instead of making the probability distribution be determined by the sum of the coefficients, we could take the sum of the squares. That is, before, we had $|v\rangle = \sum p_I|I\rangle$ with $p_I \geq 0$ and $\sum_I p_I = 1$ and the probability of getting output I was p_I. Instead we may take $\sum q_I|I\rangle$ where $\sum q_I^2 = 1$ and the probability of getting I is q_I^2. If we do this, there is no harm in allowing the entries of the output vectors to become negative, and to perform a coin flip, one could use

$$H := \frac{1}{\sqrt{2}}\begin{pmatrix} 1 & 1 \\ 1 & -1 \end{pmatrix} \tag{2.1.1}$$

applied to $|0\rangle$. The matrix H is called the *Hadamard matrix* or *Hadamard gate* in the quantum computing literature. It could just as well be called the *quantum coin flip*.

Exercise 2.1.1. We have seen the matrix H without the $\frac{1}{\sqrt{2}}$ normalization before — where?

Remark 2.1.2. One indication that the wish list above may be related to quantum mechanics is that the interference patterns observed in the famous two slit experiment (see §2.2 below) are manifested in generalized probability: a "random bit" is obtained by applying H to $|0\rangle$: $H|0\rangle = \frac{1}{\sqrt{2}}(|0\rangle + |1\rangle)$. However, if we apply a second quantum coin flip to the vector, we loose the randomness as $H^2 = \mathrm{Id}$, which, as pointed out in [**Aar13**], could be interpreted as a manifestation of interference.

The set of stochastic matrices preserve the ℓ_1-norm, $||v||_1 = \sum_I |v_I|$. The stochastic matrices must be replaced with the matrices that map vectors of length one in the ℓ_2-norm to themselves, where the ℓ_2-norm of a real-valued vector $|v\rangle = \sum v_I |I\rangle$ is defined to be

$$|v| := \sqrt{\sum v_I^2}. \tag{2.1.2}$$

These are the *orthogonal matrices*:

$$\mathbf{O}(n) := \{A \in \mathrm{Mat}_{n\times n}(\mathbb{R}) \mid |Av| = |v| \ \forall |v\rangle \in \mathbb{R}^n\}. \tag{2.1.3}$$

Exercise 2.1.3. Show that

$$\mathbf{O}(n) = \{A \in \mathrm{Mat}_{n\times n}(\mathbb{R}) \mid A^{\mathbf{t}}A = \mathrm{Id}\}.$$

With this change, we obtain our second wish as orthogonal matrices are by definition invertible (with $A^{-1} = A^{\mathbf{t}}$), and they form a group (as $(AB)^{\mathbf{t}} = B^{\mathbf{t}}A^{\mathbf{t}}$). Moreover many operations satisfy wish (3). For example, rotation matrices are orthogonal, and any rotation matrix has a square root that is a rotation matrix. (This will be a special case of the analogous result for the unitary group, Proposition 2.1.5.)

Our third property will not be completely satisfied, as the matrix

$$\begin{pmatrix} 1 & 0 \\ 0 & -1 \end{pmatrix} \in \mathbf{O}(2),$$

which represents a reflection, does not have a square root in $\mathbf{O}(2)$.

To have the third wish satisfied, we will allow ourselves vectors with *complex* entries. Recall that, except when used as an index, $i = \sqrt{-1}$. For a complex number $z = x + iy$, its complex conjugate is $\bar{z} = x - iy$, and $|z|^2 = z\bar{z}$ is the square of its norm. For a vector $|v\rangle = \sum z_I |I\rangle$, where $z_I \in \mathbb{C}$, the length of v is

$$|v| := \sqrt{\sum_I |z_I|^2}. \tag{2.1.4}$$

For $A \in \mathrm{End}(\mathbb{C}^n)$, one has $|A| := \sqrt{\sum_{j,k=1}^n |A_{jk}|^2}$.

We will work with vectors $|v\rangle = \sum z_I |I\rangle$, where $z_I \in \mathbb{C}$ and $\sum \overline{z_I} z_I = 1$. The probability of obtaining I is now $|z_I|^2$. The set of matrices preserving the norm (2.1.4) is the *unitary group*:

Definition 2.1.4. The *unitary group* of $\mathbb{C}^n$ is

$$\mathbf{U}(n) := \{A \in \mathrm{Mat}_{n\times n}(\mathbb{C}) \mid |Av| = |v| \; \forall |v\rangle \in \mathbb{C}^n\}.$$

Claim: Elements of $\mathbf{U}(n)$ satisfy the third wish on the list. More precisely:

Proposition 2.1.5. *For all $A \in \mathbf{U}(n)$, there exists a matrix $B \in \mathbf{U}(n)$ satisfying $B^2 = A$. Moreover, fix any $\epsilon > 0$. Then given any $A \in \mathbf{U}(n)$, there exists a natural number $M = M(\epsilon, n)$ and a unitary matrix $C \in \mathbf{U}(n)$ such that $|\operatorname{Id} - C| < \epsilon$ and $A = C^M$.*

Proposition 2.1.5 is proved in §2.1.2 below.

Proposition 2.1.5 is important for computation: recall that if generalized probability can be implemented, it would need to be implemented with a finite gate set from elements of $\mathbf{U}(n)$. We would want to be able to approximate any unitary matrix using a small set of such elements. Proposition 2.1.5 is a step in that direction as it reduces the problem of approximating an arbitrary element to approximating elements close to the identity.

Consider wish (1): does allowing unitary matrices increase computational power? It is an open question! However, we can at least see that our generalized probabilistic computation includes our old probabilistic computation by the following easy exercise:

Exercise 2.1.6. Show that quantum coin flip H, the not ($\neg$) matrix, and the Toffoli matrix (1.5.1) are unitary.

Moreover, we will see quantum algorithms that are better than any *known* probabilistic algorithms.

So we have gone from bits

$$\{|0\rangle, |1\rangle\}$$

to pbits

$$\{p|0\rangle + q|1\rangle \mid p, q \geq 0 \text{ and } p + q = 1\}$$

to *qubits*

$$\{\alpha|0\rangle + \beta|1\rangle \mid \alpha, \beta \in \mathbb{C} \text{ and } |\alpha|^2 + |\beta|^2 = 1\}. \tag{2.1.5}$$

The set (2.1.5), considered in terms of real parameters, is the 3-sphere S^3 in $\mathbb{R}^4 \simeq \mathbb{C}^2$. However, the probability distributions induced by $|\psi\rangle$ and $e^{i\theta}|\psi\rangle$ are the same, so the set of qubits is really S^3/S^1 (the Hopf fibration), which is the 2-sphere S^2. Physicists call this S^2 the “Bloch sphere”.

Remark 2.1.7. While we follow standard conventions from physics, from a geometric perspective it would be more natural to work with projective space $\mathbb{CP}^1 \simeq S^2$ as our space of qubits, instead of a subset of $\mathbb{C}^2$.

The norm (2.1.4) induces a Hermitian inner product via the norms of v, w, $v + w$, and $v + iw$ as follows:

$$\begin{aligned}\langle v|w\rangle : &= \overline{v_1}w_1 + \cdots + \overline{v_n}w_n \\ &= \frac{1}{2}[|v + w|^2 - i|v + iw|^2 - (1 - i)|v|^2 - (1 - i)|w|^2].\end{aligned}$$

Here and throughout this book we follow the physics convention of having the product be conjugate linear in the first factor and linear in the second, which is the reverse of the mathematics convention.

Introduce the notation that for a column vector $|w\rangle = (w_1, \dots, w_n)^{\mathbf{t}}$, the row vector

$$\langle w| := (\overline{w_1}, \dots, \overline{w_n}), \tag{2.1.6}$$

so $\langle w|v\rangle = \sum_j \overline{w_j} v_j$ is row-column matrix multiplication.

2.1.2. Exercises on the unitary group and proof of Proposition 2.1.5.

The unitary group will play a central role in all that follows. Here are some basic exercises.

Exercise 2.1.8. Show that if $A \in \mathbf{U}(n)$, then $\langle v|w\rangle = \langle Av|Aw\rangle$ for all $v, w \in \mathbb{C}^n$.

Exercise 2.1.9. Show that $\mathbf{U}(n) = \{A \in \mathrm{Mat}_{n\times n}(\mathbb{C}) \mid \bar{A}^{\mathbf{t}}A = \mathrm{Id}\}$.

Exercise 2.1.10. Let $A \in \mathbf{U}(n)$ and let λ be an eigenvalue of A. Show that $|\lambda| = 1$. In particular, we may write $\lambda = e^{i\theta}$ for some $\theta \in [0, 2\pi)$.

Exercise 2.1.11. Let $A \in \mathbf{U}(n)$ and let λ be an eigenvalue of A with eigenvector $|v\rangle$. Set $v^\perp := \{|w\rangle \in \mathbb{C}^n \mid \langle w|v\rangle = 0\}$, the orthogonal complement to $|v\rangle$ in $\mathbb{C}^n$. Show that $A(v^\perp) = v^\perp$.

Exercise 2.1.12. Let $A \in \mathbf{U}(n)$. Show that $\mathbb{C}^n$ has a basis $|v_1\rangle, \dots, |v_n\rangle$ of eigenvectors for A that is orthonormal; i.e., $\langle v_j|v_k\rangle = \delta_{jk}$ for all j, k. ⊙

A matrix is diagonalizable if and only if it admits a basis of eigenvectors. Thus Exercise 2.1.12 shows that a unitary matrix is diagonalizable.

Proof of Proposition 2.1.5. Let $|v_1\rangle, \dots, |v_n\rangle$ be an eigenbasis of $\mathbb{C}^n$ for $A \in \mathbf{U}(n)$ where $|v_j\rangle$ has eigenvalue $e^{i\theta_j}$, with $-\pi < \theta_j \le \pi$. Let $B : \mathbb{C}^n \to \mathbb{C}^n$ be the matrix with the property that $B|v_j\rangle = e^{i\frac{\theta_j}{2}}|v_j\rangle$. By construction $B^2 = A$. Moreover, B preserves the lengths of the eigenvectors. Since the eigenvectors $|v_1\rangle, \dots, |v_n\rangle$ form an orthogonal basis by Exercise 2.1.12, B preserves the lengths of all vectors and is therefore unitary. The "Moreover" assertion is similar: take C such that $C|v_j\rangle = e^{i\frac{\theta_j}{M}}|v_j\rangle$, where

$$M > \frac{\pi\sqrt{2n}}{\epsilon}.$$

Write $\theta = \max_j |\theta_j|$ and recall $\theta \leq \pi$. We have

$$\begin{aligned}
&|\operatorname{Id} - C|^2\\
&< n(1 - e^{i\theta/M})(1 - e^{-i\theta/M})\\
&= n(1 - 2\cos(\theta/M) + 1)\\
&= n\left(1 - 2\left(1 - \frac{1}{2}\frac{\theta^2}{M^2} + \frac{1}{4!}\frac{\theta^4}{M^4} - O\left(\frac{1}{M^6}\right)\right) + 1\right)\\
&< n\frac{\theta^2}{M^2} + O\left(\frac{1}{M^4}\right).
\end{aligned}$$

Since $n\frac{\theta^2}{M^2} < \frac{\epsilon^2}{2} \ll \epsilon$ as long as ϵ is sufficiently small, we conclude. □

Exercise 2.1.13. Show that the matrix $\frac{1}{\sqrt{N}}\mathrm{DFT}_N$ from §1.2.2 is unitary.

2.1.3. Hilbert spaces.

Although the Hilbert spaces in this book will always be isomorphic to some $\mathbb{C}^n$ equipped with the norm (2.1.4), there will not in general be any canonical isomorphism. Thus it will be useful to have a definition of a Hilbert space in basis-free language.

One example where it is not advantageous to use column vectors is as follows: if one writes spaces of linear maps and tensors as column vectors, it is hard to keep track of the extra structure they have as linear maps or tensors.

Definition 2.1.14. A (finite-dimensional) *Hilbert space* $\mathcal{H}$ is a (finite-dimensional) complex vector space equipped with a nondegenerate Hermitian inner-product,

$$h : \mathcal{H} \times \mathcal{H} \to \mathbb{C},$$

where h is conjugate linear in the first factor and linear in the second, $h(|v\rangle, |w\rangle) = \overline{h(|w\rangle, |v\rangle)}$, and $h(|v\rangle, |v\rangle) > 0$ for all $|v\rangle \neq 0$. As mentioned above, the conjugate linearity in the first factor is the physicists' convention.

The Hermitian inner-product h allows a canonical identification of $\mathcal{H}$ with $\mathcal{H}^*$ by

$$\begin{aligned}
h_1 &: \mathcal{H} \to \mathcal{H}^* \\
&|w\rangle \mapsto \langle w| := h(|w\rangle, \cdot).
\end{aligned} \tag{2.1.7}$$

Note that the map h_1 is *not* complex linear. This identification will be used repeatedly. Write $h(|v\rangle, |w\rangle) = \langle v|w\rangle$ and $|v| = \sqrt{\langle v|v\rangle}$ for the length of $|v\rangle$.

If $\mathcal{H} = \mathbb{C}^n$ with its standard basis, where $|v\rangle = (v_1, \dots, v_n)^{\mathbf{t}}$, the *standard Hermitian inner-product* on $\mathbb{C}^n$ is $\langle w|v\rangle = \sum_j \bar{w}_j v_j$. We always assume $\mathbb{C}^n$ is equipped with its standard Hermitian inner-product.

Definition 2.1.15. The *adjoint* of an operator $X \in \mathrm{End}(\mathcal{H})$, denoted $X^\dagger \in \mathrm{End}(\mathcal{H})$, is the operator such that for all $|v\rangle, |w\rangle \in \mathcal{H}$,

$$h(|X^\dagger v\rangle, |w\rangle) = h(|v\rangle, |Xw\rangle).$$

Call X *Hermitian* if $X = X^\dagger$. Let $\mathrm{Herm}(\mathcal{H}) \subset \mathrm{End}(\mathcal{H})$ denote the Hermitian endomorphisms of $\mathcal{H}$.

More generally, for $X \in \mathrm{Hom}(\mathcal{H}_1, \mathcal{H}_2)$, define $X^\dagger \in \mathrm{Hom}(\mathcal{H}_2, \mathcal{H}_1)$ by, for all $|v_1\rangle \in \mathcal{H}_1$, $|w_2\rangle \in \mathcal{H}_2$,

$$h_{\mathcal{H}_1}(X^\dagger|w_2\rangle, |v_1\rangle) = h_{\mathcal{H}_2}(|w_2\rangle, X|v_1\rangle).$$

Write $h_1(|Xv\rangle) = \langle v|X^\dagger$.

Definition 2.1.16. For a Hilbert space $\mathcal{H}$, define its *unitary group*

$$\mathbf{U}(\mathcal{H}) := \{U \in \mathrm{End}(\mathcal{H}) \mid |Uv| = |v| \ \forall |v\rangle \in \mathcal{H}\}.$$

When $\mathcal{H} = \mathbb{C}^n$, $\mathbf{U}(\mathbb{C}^n) = \mathbf{U}(n)$.

Exercise 2.1.17. Show that

$$\begin{aligned}\mathbf{U}(\mathcal{H}) &= \{U \in \mathrm{End}(\mathcal{H}) \mid \langle Uv|Uw\rangle = \langle v|w\rangle \ \forall |v\rangle, |w\rangle \in \mathcal{H}\}\\ &= \{U \in \mathrm{End}(\mathcal{H}) \mid U^\dagger U = \mathrm{Id}\}.\end{aligned}$$

Exercise 2.1.18. Show that if $\mathcal{H} = \mathbb{C}^n$, then $X^\dagger = \bar{X}^{\mathbf{t}}$.

Exercise 2.1.19. Show that the eigenvalues of a Hermitian matrix are real.

Exercise 2.1.20. Show that if $Y \in \mathrm{End}(\mathcal{H})$ is arbitrary, then $YY^\dagger$ and $Y^\dagger Y$ are Hermitian.

2.2. Experimental origins of quantum mechanics

For those familiar with the mathematics of electromagnetism (e.g., Maxwell's equations) or even just water waves, the difference with the mathematics of quantum physics is not so vast. However, there were phenomena observed at the beginning of the twentieth century that were not explained by existing physical theories. We describe one, in simplified form. The reader is encouraged to read Chapter 1 of [**FLS65**] for more details.

We have already mentioned "interference" in Remark 2.1.2, but this is actually a classical phenomenon observable in water waves. If traveling water waves hit a wall with a gap in it, as they emerge from the gap they will disperse in a circular pattern. If there are two gaps, the patterns interfere with each other, forming diffraction patterns.

Compare this with marbles being sent out from a source in all directions. They will spread out when they emerge from a gap independently of whether there is one gap or two. Thus there is no interference.

When one performs a similar experiment with photons or electrons, the following occurs: if one "keeps track" of the photons passing through a pair of slits, the behavior is like the marbles, but if one does not, there is interference as with waves.

The "keeping track" of the photons may be made precise using Feynman's path formalism. See [**Fey85**] for a general audience discussion.

These and other observations led to the "particle-wave" description of nature and the postulates of quantum mechanics.

In order to keep the technical difficulties to a minimum, throughout this book we restrict our attention to finite-dimensional subspaces of the infinite-dimensional Hilbert space of quantum mechanics. This is a typical approximation in the quantum information context, where the lowest eigenstates of the systems under consideration are well separated in energy from the higher excited ones.

2.3. Postulates of quantum mechanics

What follows are the standard postulates of quantum mechanics. We will show that they are essentially the same as our notion of generalized probability. The reader is encouraged to consult a physics text such as [**FLS65**] for proper physics motivation. The word "postulate" is used here in the physics sense, not the mathematical sense, in that the postulates are extracted from experimental evidence.

2.3.1. Postulate 1: State space. The first postulate describes the space one works in:

P1. Associated to any isolated physical system is a Hilbert space $\mathcal{H}$, called the *state space*. The system is completely described at a given moment by a unit vector $|\psi\rangle \in \mathcal{H}$, called its *state vector*, which is initially well-defined up to a phase $e^{i\theta}$ with $\theta \in \mathbb{R}$.

Remark 2.3.1. The first postulate is identical to what one gets with generalized probability.

2.3.2. Postulate 2: Evolution. The second postulate describes how a state vector evolves over time:

P2. The state of an isolated system evolves over time according to the *Schrödinger equation*

$$i\hbar\frac{d|\psi\rangle}{dt} = X|\psi\rangle,$$

where $\hbar$ is a constant, called *Planck's constant*, and X is a fixed Hermitian operator, called the *Hamiltonian* of the system.

Remark 2.3.2. Physicists tend to use the letter H for the Hamiltonian, but since H is already being used for the Hadamard matrix and $\mathcal{H}$ is used for Hilbert spaces, we do not adopt this convention.

Remark 2.3.3. While the phase discussed in Postulate 1 at time zero is arbitrary, it fixes the phase for $t > 0$.

2.3.3. Relation of Postulate 2 to generalized probability. Recall that in generalized probability theory, transformations are unitary.

How are unitary operators from generalized probability related to Schrödinger's equation? The answer will be given by equation (2.3.2). We first study curves in $\mathbf{U}(\mathcal{H})$. Let $U(t) \subset \mathbf{U}(\mathcal{H})$ be a smooth curve with $U(0) = \text{Id}$. Write $U'(0) = \frac{d}{dt}|_{t=0}U(t)$. Consider, for fixed vectors $|v\rangle, |w\rangle \in \mathcal{H}$,

$$\begin{aligned} 0 &= \frac{d}{dt}|_{t=0}h(|v\rangle, |w\rangle) \\ &= \frac{d}{dt}|_{t=0}\langle v|U(t)^{\dagger}U(t)w\rangle \\ &= \langle v|U'(0)^{\dagger}w\rangle + \langle v|U'(0)w\rangle. \end{aligned} \tag{2.3.1}$$

Since this occurs for all v, w, we obtain $U'(0)^{\dagger} = -U'(0)$. Compare this with a Hermitian operator X which satisfies $X^{\dagger} = X$.

Remark 2.3.4. The trick of writing 0 as the derivative of a constant function is ubiquitous in differential geometry.

Exercise 2.3.5. Show that $iU'(0)$ is Hermitian.

Definition 2.3.6. Let $\mathcal{H}$ be a Hilbert space. Define the *unitary Lie algebra*

$$\mathfrak{u}(\mathcal{H}) := \{X \in \text{End}(\mathcal{H}) \mid X^{\dagger} = -X\}.$$

Recall that when we make an identification $\mathcal{H} \cong \mathbb{C}^n$, we have $X^{\dagger} = \bar{X}^{\mathbf{t}}$. Thus $\mathfrak{u}(\mathcal{H})$ is a *real* vector space, not a complex one, because complex conjugation is not a complex linear map. As a real vector space $\text{End}(\mathbb{C}^n) = \mathbb{R}^{2n} \otimes \mathbb{R}^{2n*}$ and $\mathfrak{u}(\mathbb{C}^n)$ is a linear subspace of this real vector space.

For $X, Y \in \text{End}(\mathcal{H})$, let

$$[X, Y] := XY - YX \in \text{End}(\mathcal{H})$$

denote their commutator.

Exercise 2.3.7. For $X, Y \in \mathfrak{u}(\mathcal{H})$ show that $[X, Y] \in \mathfrak{u}(\mathcal{H})$, showing that $\mathfrak{u}(\mathcal{H})$ is indeed a Lie algebra (see Definition A.2.4) with the multiplication given by the commutator.

Definition 2.3.8. For a smooth subset Σ of a vector space V and for $x \in \Sigma$, the *tangent space* to Σ at x, denoted $T_x\Sigma$, is the span of all tangent vectors $c'(0)$ to smooth curves $c(t) \subset \Sigma$ with $c(0) = x$.

By (2.3.1),

$$\mathfrak{u}(\mathcal{H}) = T_{\mathrm{Id}}\mathbf{U}(\mathcal{H}),$$

the tangent space to the unitary group at the identity.

Write $\mathfrak{u}(n) = \mathfrak{u}(\mathbb{C}^n)$. We have $\mathfrak{u}(n) = \{X \in \mathrm{End}(\mathbb{C}^n) \mid \bar{X}^{\mathbf{t}} = -X\}$. Thus $i\mathfrak{u}(\mathcal{H}) \subset \mathrm{End}(\mathcal{H})$ is the space of Hermitian endomorphisms.

Exercise 2.3.9. Verify independently that both $\mathfrak{u}(n)$ and the space of $n \times n$ Hermitian matrices have (real) dimension n^2.

The space $\mathfrak{u}(\mathcal{H})$ is also called the *Lie algebra of* $\mathbf{U}(\mathcal{H})$.

Aside 2.3.10. For those familiar with dimensions of manifolds, Exercise 2.3.9 also shows $\dim \mathbf{U}(n) = n^2$.

Exercise 2.3.11. Show that elements of $\mathfrak{u}(\mathcal{H})$ have purely imaginary eigenvalues. ⊙

Exercise 2.3.12. Show that elements of $\mathfrak{u}(\mathcal{H})$ admit a basis of eigenvectors. ⊙

We next show how to go from elements of the Lie algebra to elements of the Lie group.

For $X \in \mathrm{End}(\mathcal{H})$, write $X^k \in \mathrm{End}(\mathcal{H})$ for $X \cdots X$ applied k times. Write

$$e^X := \sum_{k=0}^{\infty} \frac{1}{k!} X^k.$$

This sum converges to a fixed endomorphism, essentially for the same reason it does in the $\dim \mathcal{H} = 1$ case.

Exercise 2.3.13. Show that the sum indeed converges, assuming the scalar case. ⊙

Proposition 2.3.14. *If X is Hermitian, then $e^{iX} \in \mathbf{U}(\mathcal{H})$.*

Exercise 2.3.15. Prove Proposition 2.3.14. ⊙

Proposition 2.3.14 shows that Postulate 2 is compatible with generalized probability:

Write $U(t) = e^{\frac{-itX}{\hbar}}$ and $|\psi_t\rangle = U(t)|\psi_0\rangle$. Then

$$\begin{aligned} \frac{d|\psi_t\rangle}{dt} &= U'(t)|\psi_0\rangle \\ &= \frac{-iX}{\hbar} e^{\frac{-itX}{\hbar}} |\psi_0\rangle \\ &= \frac{-iX}{\hbar} |\psi_t\rangle, \end{aligned} \tag{2.3.2}$$

which is Schrödinger's equation.

2.3.4. Postulate 3: Measurements. Quantum measurement is the process through which one extracts classical information out of a quantum system.

In our first two postulates we dealt with isolated systems. In reality, no system is isolated and the whole universe is modeled by one enormous Hilbert space. In practice, parts of the system are sufficiently isolated that they can be treated as isolated systems. However, they are occasionally acted upon by the outside world, and we need a way to describe the action of the outside world on the otherwise isolated system.

For our purposes, the isolated systems will be finite-dimensional Hilbert spaces attached to the inputs of a quantum algorithm and the outside interference will be the measurement at the end. After a sequence of unitary operations one obtains a vector $|\psi\rangle = \sum z_j|j\rangle$ of norm one, and as in generalized probability:

P3. If $|\psi\rangle = \sum_j z_j|j\rangle$ and a measurement is taken, the output is j with probability $|z_j|^2$. If j_0 is the output, after the measurement the state becomes $|j_0\rangle$.

Remark 2.3.16. In §5.1 more general notions of measurement are introduced.

2.3.5. Postulate 4: Composite systems. A typical situation in quantum information theory is that there are two or more isolated systems, say $\mathcal{H}_A$, $\mathcal{H}_B$, that are brought together (i.e., allowed to interact with each other) to form a larger isolated system $\mathcal{H}_{AB}$. The larger system is called the *composite system.*

P4. The state space of a composite system $\mathcal{H}_{AB}$ is the tensor product of the state spaces of the component physical systems $\mathcal{H}_A, \mathcal{H}_B$:

$$\mathcal{H}_{AB} = \mathcal{H}_A \otimes \mathcal{H}_B.$$

Postulate 4 is the same as what one gets with generalized probability.

The Hermitian inner-product on $\mathcal{H}_A \otimes \mathcal{H}_B$ is inherited from those on $\mathcal{H}_A$, $\mathcal{H}_B$:

$$h_{AB}(|a\rangle\otimes|b\rangle, |a'\rangle\otimes|b'\rangle) = h_A(|a\rangle, |a'\rangle)h_B(|b\rangle, |b'\rangle).$$

Remark 2.3.17. In §5.2.4, we will discuss how one may *derive* the postulates of quantum mechanics as a natural generalization of probability by just insisting on allowing operations arbitrarily close to the identity. The perspective of quantum mechanics as generalized probability now has a considerable literature and has given rise to a notion of "general probabilistic theories"; see **[Plá]** for a survey.

2.3.6. Entanglement. This tensor product structure gives rise to the notion of *entanglement*, which, as we will see in the next few sections, gives rise to phenomena outside of our classical intuition.

Definition 2.3.18. A state $|\psi\rangle \in \mathcal{H}_1 \otimes \cdots \otimes \mathcal{H}_n$ is called *separable* if $|\psi\rangle = |v_1\rangle \otimes \cdots \otimes |v_n\rangle$ with each $|v_j\rangle \in \mathcal{H}_j$. Otherwise it is *entangled*.

The terminology *rank one tensor* is used for separable states in the mathematics literature.

Physicists describe their experiments in terms of two characters, Alice and Bob. We will generally follow this convention and name Hilbert spaces after them.

The strange thing about entanglement is that it persists over distances. This was experimentally verified in 1972 by Fry and Thompson **[FT76]** and more definitively the 1980s by A. Aspect and his group **[A99]** for which he shared the 2022 Nobel Prize in Physics. The experiment verified the set-up described in §2.6: an atom decays and emits a pair of photons. One goes towards Alice's region (described by Alice's Hilbert space $\mathcal{H}_A$), the other towards Bob's region (described by Bob's Hilbert space $\mathcal{H}_B$). Entanglement is generated by the emission process; e.g., let $\mathcal{H} = \mathbb{C}^2 \otimes \mathbb{C}^2 = \mathcal{H}_A \otimes \mathcal{H}_B$, and let

$$|\mathrm{epr}\rangle := \frac{|00\rangle + |11\rangle}{\sqrt{2}} \tag{2.3.3}$$

denote a state generated by the emission process. This state $|\mathrm{epr}\rangle$ is called the *EPR state* in the physics literature, named after Einstein-Podolsky-Rosen.

The vector $|\mathrm{epr}\rangle$ is "stored" in the photons. At this point, Alice may operate on her photon by unitary operations and measurement, and Bob on his. In what follows we will see three examples of phenomena that go outside of our classical intuition resulting from this set-up. For example, if Alice measures her photon in the standard basis, she will obtain one of 0 or 1, each with equal probability. If Bob measures his after Alice had measured 0, Bob is measuring the classical state $|0\rangle$ and thus will get 0, and similarly he will get 1 if Alice had measured 1. So far, this could also be accounted for by classical probability. However, say Alice measures in a different orthonormal basis, say $|\psi\rangle, |\phi\rangle$. Then similarly if Alice obtains $|\psi\rangle$ (resp. $|\phi\rangle$), then if Bob measures in the basis $|\psi\rangle, |\phi\rangle$, he will also obtain $|\psi\rangle$ (resp. $|\phi\rangle$). This basis-independent correlation cannot be accounted for by classical probability. This led Einstein-Podolsky-Rosen [**EPR35**] to believe something was wrong with quantum mechanics; see §2.6.

2.4. Super-dense coding

This section discusses how, with a shared entangled state, it is possible to transmit two bits of classical information by transmitting just one qubit. This phenomenon is called "super[1]-dense coding". Super-dense coding was introduced in [**BW92**].

Before the experiment begins, Alice and Bob agree on a protocol for communication as described below. Assume they have a "quantum channel" along which they can transmit qubits, just as a classical channel transmits classical bits. (Quantum channels will be defined in §5.5.)

Assume $|\mathrm{epr}\rangle$ of (2.3.3) has been created, both Alice and Bob are aware of it, Alice is in possession of (i.e., can manipulate) the first qubit, and Bob the second, as described in the Fry/Aspect experiment above.

Now say Alice wants to transmit a two classical bit message to Bob, i.e., one of $00, 01, 10, 11$. She is allowed to act on her half of $|\mathrm{epr}\rangle$ by unitary transformations and then send it to Bob. (Later we will establish a gate set she must choose from, but it will include the gates used below.) Once Alice sends her qubit to Bob, he is in control of two qubits that he can act on locally. They decided in advance that after Alice sends her part of the qubit to Bob, he will act on the four-dimensional space $\mathcal{H}_A \otimes \mathcal{H}_B$ by the unitary operator that performs the following change of basis and then measure:

$$\begin{aligned}|\mathrm{epr}\rangle = \frac{1}{\sqrt{2}}(|00\rangle + |11\rangle) &\mapsto |00\rangle,\\ \frac{1}{\sqrt{2}}(|00\rangle - |11\rangle) &\mapsto |01\rangle,\\ \frac{1}{\sqrt{2}}(|10\rangle + |01\rangle) &\mapsto |10\rangle,\\ \frac{1}{\sqrt{2}}(|10\rangle - |01\rangle) &\mapsto |11\rangle.\end{aligned}$$

If Alice wants to send 00, she just does nothing before transmitting her qubit to Bob, so Bob will receive $|\mathrm{epr}\rangle$ and when he measures he will get $|00\rangle$ with probability one. Similarly, if she wants to send 01, before transmitting she acts on her qubit by

$$\begin{pmatrix}1 & 0\\ 0 & -1\end{pmatrix}$$

[1] Physicists are fond of the word "super".

so that Bob will be in possession of the state $\frac{1}{\sqrt{2}}(|00\rangle - |11\rangle)$. When he performs the change of basis and measures, he will get $|01\rangle$ with probability one.

Exercise 2.4.1. What are the other two matrices Alice should act by to transmit the other two-bit messages 10 and 11?

In summary, with preparation of an EPR state in advance, plus transmission of a single qubit, one can transmit two classical bits of information.

Remark 2.4.2. This rate of transmission is optimal; see [**Wat18**, §8.1.3].

2.5. Quantum teleportation

In this set-up Alice wants to "send" a qubit

$$|\psi\rangle = \alpha|0\rangle + \beta|1\rangle$$

to Bob that she is in possession of. She and Bob have a shared $|\text{epr}\rangle$ and access to a classical channel along which she may send bits to Bob. We will see that she can transmit $|\psi\rangle$ to Bob by sending him two classical bits.

Write the state of the system on $(\mathbb{C}^2 \otimes \mathbb{C}^2)_A \otimes \mathbb{C}^2_B$ as

$$\begin{aligned}
&|\psi\rangle \otimes |\text{epr}\rangle \\
&= (\alpha|0\rangle + \beta|1\rangle) \otimes \frac{1}{\sqrt{2}}(|00\rangle + |11\rangle) \\
&= \alpha|00\rangle \otimes |0\rangle + \alpha|01\rangle \otimes |1\rangle + \beta|10\rangle \otimes |0\rangle + \beta|11\rangle \otimes |1\rangle.
\end{aligned}$$

Alice can operate on the first two qubits.

Recall the CNOT gate from Exercise 1.4.2:

$$\begin{aligned}
|00\rangle &\mapsto |00\rangle, \\
|01\rangle &\mapsto |01\rangle, \\
|10\rangle &\mapsto |11\rangle, \\
|11\rangle &\mapsto |10\rangle,
\end{aligned}$$

and that the first bit controls whether NOT (i.e., $\begin{pmatrix} 0 & 1 \\ 1 & 0 \end{pmatrix}$) or Id is applied to the second bit.

Exercise 2.5.1. Show that if Alice acts on the first two qubits by a CNOT gate followed by $H \otimes \text{Id} = \frac{1}{\sqrt{2}}\begin{pmatrix} 1 & 1 \\ 1 & -1 \end{pmatrix} \otimes \text{Id}$, the resulting state is

$$\begin{aligned}
&\frac{1}{2}[|00\rangle \otimes (\alpha|0\rangle + \beta|1\rangle) + |01\rangle \otimes (\alpha|1\rangle + \beta|0\rangle) \\
&\quad + |10\rangle \otimes (\alpha|0\rangle - \beta|1\rangle) + |11\rangle \otimes (\alpha|1\rangle - \beta|0\rangle)].
\end{aligned}$$

Alice acts on her qubits as in Exercise 2.5.1. Notice that Bob's coefficient of Alice's $|00\rangle$ is the state ψ that is to be transmitted. Alice performs a measurement. If she has the good luck to obtain $|00\rangle$, then she knows Bob has $|\psi\rangle$ and she can tell him classically that he is in possession of $|\psi\rangle$. But say she obtains the state $|01\rangle$: the situation is still good, she knows Bob is in possession of a state such that if he acts on it with $\begin{pmatrix} 0 & 1 \\ 1 & 0 \end{pmatrix}$, he will obtain the state $|\psi\rangle$, so she just needs to tell him classically to apply it. Since they had communicated the algorithm in the past, all Alice really needs to tell Bob in the first case is the classical message 00 and in the second case the message 01.

Exercise 2.5.2. Write out the actions Bob should take when Alice sends him 01 and when she sends him 11.

Remark 2.5.3. The name "teleportation" is misleading because information is transmitted at a speed slower than the speed of light.

2.6. Bell's game

As mentioned above, the 1935 Einstein-Podolsky-Rosen paper [**EPR35**] challenged quantum mechanics with the following thought experiment that they believed implied that quantum mechanics was not complete and that there must be additional (hidden) variables. They argued that, otherwise, instantaneous communication across distances would occur, in violation of principles of relativity: Alice and Bob prepare $|epr\rangle = \frac{1}{\sqrt{2}}(|00\rangle + |11\rangle)$, with Alice in control of the first qubit and Bob the second, and then they travel far apart. Alice measures her qubit. Recall that if she gets 0, then she can predict with certainty that Bob will get 0 in his measurement, even if his measurement is taken a second later and they are a light-year apart.

Ironically, this thought experiment was made into a proposed physical experiment designed by Bell [**Bel64**] and carried out by E. Fry and R. Thompson [**FT76**] and by A. Aspect and others [**A99**]. Bell showed that if the experiment could be performed, the measurement phenomenon could not be accounted for by additional variables alone. Bell's argument proceeds by assuming a continuous probability distribution governs the additional variables and then shows that no such continuous probability distribution can exist. What follows is a version from [**CHSH69**], adapted from the presentation in [**AB09**] given in a discrete setting.

Remark 2.6.1. According to F. Holweck (personal communication): in the 1930s the debate between Einstein and Bohr over the completeness of quantum mechanics was not really viewed as a scientific question, but rather a philosophical debate on reality between two geniuses. After Bell, the status of the problem changed, and after the experimental verification, it changed status again, as entanglement started to be considered as a resource for quantum information/technology. This illustrates the great talent of Einstein and Bohr to recognize that philosophical questions really matter. See [**Hol**] and [**Geo21**] for further reading.

One interpretation of the result of the experiment (e.g., in [**AB09**]) is that Alice and Bob are acting on information that has been previously shared. Another standard interpretation was communicated to me by S. Aaronson: the experiment shows "that local hidden variable models are experimentally ruled out, and ALSO there's no superluminal communication, and the way to explain the result is just to accept that the world is quantum-mechanical and entanglement is real."

A game is used to show that neither classical nor probabilistic reasoning can account for the experimental results. Alice and Bob are on the same team and Charlie is a referee. Charlie chooses $x, y \in \{0, 1\}$ by flipping a fair coin and sends x to Alice and y to Bob. Based on this information, Alice and Bob, without communicating with each other, get to choose bits a, b and send them to Charlie. They win if $a \oplus b = x \wedge y$, i.e., either $(x, y) \neq (1, 1)$ and $a = b$ or $(x, y) = (1, 1)$ and $a \neq b$.

2.6.1. Classical version. Note that if Alice and Bob both always choose 0, they win with probability $\frac{3}{4}$.

Theorem 2.6.2 ([**Bel64**]). *Regardless of the classical or probabilistic strategy Alice and Bob use, they never win with probability greater than* $\frac{3}{4}$.

The idea of the proof is that one first reduces a probabilistic strategy to a classical one, because after repeated rounds of the game, one can just adopt the most frequent choice for each possible x, y. Then there are only 2^4 possible strategies and each can be analyzed. See, e.g., [**AB09**, Thm. 20.2] for more detail.

2.6.2. Quantum version. Alice and Bob prepare $|\text{epr}\rangle = \frac{1}{\sqrt{2}}(|00\rangle + |11\rangle)$ in advance. Alice takes the first qubit and Bob the second. When Alice gets x from Charlie, if $x = 1$, she applies a rotation by $\frac{\pi}{8}$ to her qubit, and if $x = 0$, she does nothing. When Bob gets y from Charlie, he applies a rotation by $-\frac{\pi}{8}$ to his qubit if $y = 1$, and if $y = 0$, he does nothing. The order in which these rotations are applied does not matter because the operators on $\mathcal{H}_A \otimes \mathcal{H}_B = \mathbb{C}^2 \otimes \mathbb{C}^2$ commute. Both of them measure their respective qubits, again, the order will not matter, and they send the values obtained to Charlie.

Theorem 2.6.3. *With this strategy, Alice and Bob win with probability greater than* $\frac{4}{5}$.

Proof. If $(x, y) = (0, 0)$, then they are measuring $|\text{epr}\rangle$, so the measurement either yields 0 for both or 1 for both and $0 \oplus 0 = 1 \oplus 1 = 0 = 0 \wedge 0$, so they always win in this case.

If $(x, y) = (1, 0)$, then they are measuring

$$\frac{1}{\sqrt{2}}\left(\cos\left(\frac{\pi}{8}\right)|00\rangle + \sin\left(\frac{\pi}{8}\right)|10\rangle - \sin\left(\frac{\pi}{8}\right)|01\rangle + \cos\left(\frac{\pi}{8}\right)|11\rangle\right),$$

and the outputs are equal with probability $\frac{1}{2}\cos^2(\frac{\pi}{8}) + \frac{1}{2}\cos^2(\frac{\pi}{8}) > \frac{17}{20}$, and similarly if $(x, y) = (0, 1)$. (Note that up to this point, we are doing worse than the naïve strategy of Alice and Bob always choosing 0.)

If $(x, y) = (1, 1)$, then they are measuring

$$\begin{aligned}\frac{1}{\sqrt{2}}\Big[&\cos\left(\frac{\pi}{8}\right)\left(\cos\left(-\frac{\pi}{8}\right)|00\rangle + \sin\left(-\frac{\pi}{8}\right)|01\rangle\right)\\ &+ \sin\left(\frac{\pi}{8}\right)\left(\cos\left(-\frac{\pi}{8}\right)|10\rangle + \sin\left(-\frac{\pi}{8}\right)|11\rangle\right)\\ &- \sin\left(\frac{\pi}{8}\right)\left(-\sin\left(-\frac{\pi}{8}\right)|00\rangle + \cos\left(-\frac{\pi}{8}\right)|01\rangle\right)\\ &+ \cos\left(\frac{\pi}{8}\right)\left(-\sin\left(-\frac{\pi}{8}\right)|10\rangle + \cos\left(-\frac{\pi}{8}\right)|11\rangle\right)\Big].\end{aligned} \tag{2.6.1}$$

Use the trigonometric identities $\cos(2\alpha) = \cos^2(\alpha) - \sin^2(\alpha)$ and $\sin(2\alpha) = 2\sin(\alpha)\cos(\alpha)$ to simplify (2.6.1) to

$$\frac{1}{2}[|00\rangle - |01\rangle + |10\rangle + |11\rangle].$$

Thus they win with probability $\frac{1}{2}$, as the probability of drawing (01) or (10) is each $\frac{1}{4}$.

We see that the overall chance of winning is greater than $\frac{1}{4}(1) + \frac{1}{4}(\frac{17}{20}) + \frac{1}{4}(\frac{17}{20}) + \frac{1}{4}(\frac{1}{2}) = \frac{4}{5}$. □

Exercise 2.6.4. Show that this strategy can be improved. What is its limit? ⊚

Chapter 3

Algorithms

This chapter covers the basics of quantum computing and the standard quantum algorithms. It presents the algorithms of Grover and Simon, respectively, in §3.1 and §3.2. Elementary number theory needed for the algorithms of Miller-Rabin and Shor is presented in §3.3. In §3.4 a probabilistic algorithm, the Miller-Rabin primality test, is presented, as ideas from its proof appear in Shor's algorithm. In §3.5, we discuss admissible quantum gates. After considerable preliminaries, we present Shor's algorithm in §3.6.

3.1. Grover's search algorithm

The brute-force search problem. We have N boxes, and exactly one of them has a blue marble in it. We are allowed to open a box at unit cost. Find the blue marble.

Do you see any strategy for minimizing the cost to find the blue marble beyond brute-force search? At this writing, none is known, and it is generally conjectured there is none. With brute-force search on the average one needs to open $N/2$ boxes to find the blue marble.

Grover's quantum algorithm improves upon brute-force search, lowering the average cost from $N/2$ to a constant times $\log(N)\sqrt{N}$. (This assumes the quantum gates in the algorithm each have unit cost.) In what follows we have $N = 2^n$

Grover's Problem. Given $F_n : \mathbb{F}_2^n \to \mathbb{F}_2$, computable by a poly$(n)$-size classical circuit, find $a \in \mathbb{F}_2^n$ such that $F_n(a) = 1$ if such a exists.

A brute-force search requires a circuit of size $\mathrm{poly}(n)2^n$, as one would need to check at least half the points to find a with probability at least $\frac{1}{2}$. No classical or probabilistic algorithm is known that does better than $\mathrm{poly}(n)2^n$.

Grover [**Gro96**] found a quantum circuit of size $\mathrm{poly}(n)2^{\frac{n}{2}}$ that solves this problem with high probability. Note that it also gives a size $\mathrm{poly}(n)2^{\frac{n}{2}}$ probabilistic solution to the **NP**-complete problem SAT discussed in §1.7; the algorithm is stronger, as it not only determines existence of a solution but finds it.

We present the algorithm for the following simplified version where one is promised that there exists exactly one solution. All essential ideas of the general case are here.

The Problem. Given $F_n : \mathbb{F}_2^n \to \mathbb{F}_2$, computable by a $\mathrm{poly}(n)$-size classical circuit, and the information that $F_n(x) = 1$ has exactly one solution $x = a$, find a.

The idea of the algorithm is to start with a vector equidistant from all possible solutions and then to incrementally rotate it towards a. What is strange for our classical intuition is that we will be able to rotate towards the solution without knowing what it is, and similarly, we won't "see" the rotation matrix either.

We work in $(\mathbb{C}^2)^{\otimes n+1+s}$ where $s = s(n)$ is the number of classical bits needed to compute F_n in a reversible classical circuit, which is also the size of the classical circuit needed to compute F_n. We suppress reference to the s qubits in what follows.

The first step is to construct such a starting vector.

The following vector is the average of all the classical (observable) states:

$$|\mathrm{av}\rangle := \frac{1}{2^{\frac{n}{2}}} \sum_{I \in \{0,1\}^n} |I\rangle. \tag{3.1.1}$$

To prepare $|\mathrm{av}\rangle$, note that $H|0\rangle = \frac{1}{\sqrt{2}}(|0\rangle + |1\rangle)$. Applying $H^{\otimes n}$ to $|0\cdots 0\rangle$ transforms it to $|\mathrm{av}\rangle$. By Exercise 1.5.2, this is applying $\mathrm{DFT}_{(\mathbb{Z}/2\mathbb{Z})^n}$. The cost of this is n gates, as $H^{\otimes n}$ is the composition of

$$H \otimes \mathrm{Id}_{2,\dots,n},\ \mathrm{Id}_1 \otimes H \otimes \mathrm{Id}_{3,\dots,n},\ \dots\ ,\ \mathrm{Id}_{1,\dots,n-1} \otimes H.$$

Since $|\mathrm{av}\rangle$ is equidistant from all possible solution vectors, we have $\langle \mathrm{av}|a\rangle = \frac{1}{2^{\frac{n}{2}}}$. We want to rotate $|\mathrm{av}\rangle$ towards the unknown a.

Exercise 3.1.1. Show that $\mathbf{U}(2)$ acts transitively on lines through the origin in $\mathbb{C}^2$: i.e, given any nonzero $v, w \in \mathbb{C}^2$, there exists $U \in \mathbf{U}(2)$ such that $U|v\rangle = \lambda|w\rangle$ for some $\lambda \in \mathbb{C}^*$. ⊚

Definition 3.1.2. The *reflection* across a hyperplane $\mathcal{M} = \mathbb{C}^{m-1} \subset \mathbb{C}^m$ is the linear map that, writing $|v\rangle \in \mathbb{C}^m$ as $|v\rangle = |v_1\rangle + |v_2\rangle$ with $|v_1\rangle \in \mathcal{M}$ and $|v_2\rangle \perp \mathcal{M}$ (such an expression is unique), sends $|v\rangle \mapsto |v_1\rangle - |v_2\rangle$.

Exercise 3.1.3. Show that $\mathbf{U}(n)$ contains the reflections.

Exercise 3.1.4. Show that the product of two reflections is a rotation. More precisely, show that if $|v\rangle, |w\rangle$ are vectors in $\mathbb{R}^n$, the composition of a reflection in the hyperplane perpendicular to $|v\rangle$, followed by a reflection in the hyperplane perpendicular to $|w\rangle$, acts as a rotation in the $|v\rangle, |w\rangle$ plane by an angle equal to twice the angle between $|v\rangle$ and $|w\rangle$, and acts as the identity in the orthogonal complement to the two-plane. We may adopt the same terminology in $\mathbb{C}^n$ if for $|v\rangle, |w\rangle \in \mathbb{C}^n$, we *define* the angle θ between them by $\cos(\theta) := \frac{\langle v|w\rangle}{|v||w|}$.

In order to perform the rotation R that moves $|\mathrm{av}\rangle$ towards $|a\rangle$, we will first reflect across the hyperplane orthogonal to $|a\rangle$. In particular we will send $|a\rangle$ to $-|a\rangle$, and then we will reflect across the hyperplane orthogonal to $|\mathrm{av}\rangle$. The codimension two space orthogonal to $\mathrm{span}\{|a\rangle, |\mathrm{av}\rangle\}$ is unmoved after these two reflections.

To perform the first reflection, consider the map

$$\begin{aligned} (\mathbb{C}^2)^{\otimes n} \otimes \mathbb{C}^2 &\to (\mathbb{C}^2)^{\otimes n} \otimes \mathbb{C}^2 \\ |x\rangle \otimes |y\rangle &\mapsto |x\rangle \otimes |(y \oplus F(x))\rangle \end{aligned} \tag{3.1.2}$$

defined on basis vectors and extended linearly. The notation $\oplus$ denotes addition in $\mathbb{F}_2$. We have slightly abused notation as (3.1.2) is really the composition of $s+1$ admissible classical gates.

To execute the map (3.1.2), use the s workspace bits corresponding to y to realize s reversible classical gates. We initially set $y = 0$ so that the image is $|x0\rangle$ for $x \neq a$, and $|x1\rangle$ when $x = a$. At this point our vector is

$$\frac{1}{2^{\frac{n}{2}}} \left(\sum_{I \neq a} |I\rangle \otimes |0\rangle + |a\rangle \otimes |1\rangle \right).$$

Next apply the quantum gate $\mathrm{Id} \otimes \begin{pmatrix} 1 & 0 \\ 0 & -1 \end{pmatrix}$ which sends $|x0\rangle \mapsto |x0\rangle$ and $|x1\rangle \mapsto -|x1\rangle$. Finally apply the map $|xy\rangle \mapsto |x(y \oplus F(x))\rangle$ again.

Thus $|a0\rangle \mapsto -|a0\rangle$ and all other basis vectors $|I0\rangle$ are mapped to themselves, as desired. Our workspace bit has been set back to 0 so we may reuse it. That is, we now have the vector

$$\frac{1}{2^{\frac{n}{2}}} \left(\sum_{I \neq a} |I\rangle \otimes |0\rangle - |a\rangle \otimes |0\rangle \right).$$

Next we need to reflect across the hyperplane perpendicular to $|\text{av}\rangle$. It is easy to reflect around a basis vector, so first perform the map $H^{\otimes n}$ that sends $|\text{av}\rangle$ to $|0\cdots 0\rangle$ (recall that $H = H^{-1}$). Next reflect in the hyperplane perpendicular to $|0\cdots 0\rangle$ using the Boolean function $g : \mathbb{F}_2^n \to \mathbb{F}_2$ that outputs 1 if and only if its input is $(0,\dots,0)$ in the role of F for our previous reflection. Finally apply $H^{\otimes n}$ again so the resulting reflection is in the hyperplane perpendicular to $|\text{av}\rangle$.

The composition of these two reflections is the desired rotation R composed with $-\text{Id}$ in the plane $\text{span}\{|a\rangle, |\text{av}\rangle\}$. The $-\text{Id}$ will be harmless for our purposes when computing amplitudes. (One could alternatively reflect in the vector perpendicular to av in the plane.) Recall that $\cos(\angle(|v\rangle, |w\rangle)) = \frac{\langle v|w\rangle}{|v||w|}$. Write the angle between $|\text{av}\rangle$ and $|a\rangle$ as $\frac{\pi}{2} - \theta$, so $\sin(\theta) = \frac{1}{2^{\frac{n}{2}}}$.

Exercise 3.1.5. Approximately determine the probability that a measurement of $R|\text{av}\rangle$ will produce $|a\rangle$. ⊚

As mentioned above, when n is large the vector $R|\text{av}\rangle$ is not useful, but if we instead compose R with itself $O(\frac{1}{\theta})$ times, we obtain a vector much closer to $|a\rangle$.

Another way to keep track of the maps is to note that the first reflection is $\text{Id} -2|a\rangle\langle a|$ and the second is $\text{Id} -2|av\rangle\langle av|$. Then the rotation is

$$\begin{aligned}
&\text{Id} -2|a\rangle\langle a| - 2|\text{av}\rangle\langle\text{av}| + 4|\text{av}\rangle\langle\text{av}| \circ |a\rangle\langle a| \\
&= \text{Id} -2|a\rangle\langle a| - 2|\text{av}\rangle\langle\text{av}| + 4\langle\text{av}|a\rangle|\text{av}\rangle\langle a| \\
&= \text{Id} -2|a\rangle\langle a| - 2|\text{av}\rangle\langle\text{av}| + \frac{4}{2^{\frac{n}{2}}}|\text{av}\rangle\langle a|.
\end{aligned}$$

Exercise 3.1.6. Using the above expression, determine exactly the probability that a measurement of $R|\text{av}\rangle$ will produce $|a\rangle$.

Exercise 3.1.7. Show that applying the procedure $\lfloor\frac{\pi}{4}\rfloor 2^{\frac{n}{2}}$ times, one obtains a vector such that the probability of it being in state $|a\rangle$ after a measurement is close to 1. ⊚

3.2. Simon's algorithm

The Problem. Assume $f = f_n : \mathbb{F}_2^n \to \mathbb{F}_2^n$ is computable by a Boolean circuit of size polynomial in n with the following property: there exists $a \in \mathbb{F}_2^n$ satisfying for all $x, y \in \mathbb{F}_2^n$ with $x \neq y$, $f(x) = f(y)$ if and only if

$$x = y\underline{\oplus}a := (y_1 \oplus a_1, \dots, y_n \oplus a_n) \in \mathbb{F}_2^n.$$

Find a.

For simplicity of exposition, assume we know $a \neq (0,\dots,0)$ as well; i.e., f is not injective.

Remark 3.2.1. This problem is expected to require an exponentially large (in n) circuit on a classical or probabilistic computer. Consider the following variant where f is allowed to be difficult to compute, but we are handed a black box that will compute it for us at unit cost. If a and f are chosen at random subject to the condition that $f(x) = f(y)$ if and only if $x = y \oplus a$, then in classical computation one would expect to use the black box $2^{\frac{n}{2}}$ times before having any information at all; with fewer calls, one is likely to never get the same answer twice.

Simon [**Sim94**] gives a poly(n) size quantum circuit that obtains the solution to the problem and the variant in Remark 3.2.1.

Remark 3.2.2. Although this problem may look unnatural, the resulting algorithm inspired Shor's algorithm and its generalizations. It fits into a larger framework of problems that allow for an exponential quantum speedup over known probabilistic algorithms; see Remark 3.6.28.

Work in $(\mathbb{C}^2)^{\otimes 2n+s}$, where $s = s(n)$ is the number of classical bits needed to compute f in a reversible classical circuit. We suppress reference to the s workspace qubits in what follows. As with Grover's algorithm, we will construct a vector that "sees" the answer a, but we will not be able to see the vector. We will manipulate the vector to get information about the solution. With Grover's algorithm, we started with $|\mathrm{av}\rangle$ and with each round we incrementally rotated it towards the solution. Here, in each round we will restrict the search for a to a hyperplane. As before, first prepare $|\mathrm{av}\rangle = \frac{1}{2^{\frac{n}{2}}} \sum_{I \in \{0,1\}^n} |I\rangle$ by applying $H^{\otimes n}$ at a cost of n gates. Then apply the operation

$$
\begin{aligned}
&(\mathbb{C}^2)^{\otimes 2n} \to (\mathbb{C}^2)^{\otimes 2n} \\
&|x\rangle \otimes |z\rangle \mapsto |x\rangle \otimes |(\underline{z \oplus f(x)})\rangle,
\end{aligned}
\tag{3.2.1}
$$

defined on basis vectors and extended linearly, to $|\mathrm{av}\rangle \otimes |0^n\rangle$. One obtains

$$
\frac{1}{2^{\frac{n}{2}}} \sum_{x \in \{0,1\}^n} |x\rangle \otimes |f(x)\rangle.
$$

Analogous to the presentation of Grover's algorithm, (3.2.1) actually represents a sequence of admissible classical gates. Next measure the second n qubits of the register so the second n qubits will be in some classical state z_0. One obtains a vector that is a constant multiple of

$$
\sum_{\{x \mid f(x) = z_0\}} |x\rangle \otimes |z_0\rangle.
$$

Let x_0 be such that $f(x_0) = z_0$. Our sum collapses to

$$\frac{1}{\sqrt{2}}(|x_0\rangle + |x_0\underline{\oplus}a\rangle)\otimes|z_0\rangle.$$

We will manipulate this vector to obtain a vector perpendicular to $|a\rangle$.

Act again by $H^{\otimes n}$ on the first n qubits.

Adopt the following notation: for $x, y \in \mathbb{F}_2^n$, let

$$x \cdot y := \bigoplus_{j=1}^{n} x_j \wedge y_j \in \mathbb{F}_2$$

denote their inner product. (Here $\bigoplus$ is addition in $\mathbb{F}_2$.)

Exercise 3.2.3. Show that for $x \in \mathbb{F}_2^n$,

$$H^{\otimes n}|x\rangle = \frac{1}{2^{\frac{n}{2}}} \sum_{y \in \mathbb{F}_2^n} (-1)^{x \cdot y}|y\rangle.$$

We obtain a scalar multiple of the vector

$$\sum_{y \in \mathbb{F}_2^n} \left((-1)^{x_0 \cdot y} + (-1)^{(x_0 \underline{\oplus} a)\cdot y}\right)|y\rangle\otimes|z_0\rangle. \tag{3.2.2}$$

Since $(x\underline{\oplus}a) \cdot y = x \cdot y \oplus a \cdot y$, the term in the summand for any given y is nonzero if and only if $a \cdot y = 0$.

When we measure the vector, the first n qubits of the output will be some $y \in \mathbb{F}_2^n$ that is orthogonal to a with respect to the inner-product on $\mathbb{F}_2^n$ and is uniformly at random among such as the coefficients for each y orthogonal to a are the same. If we run the algorithm several times, with good luck, we could find a after $n-1$ runs of the algorithm. However, we have no guarantee we do not end up with linearly dependent y's.

Exercise 3.2.4. Prove that for any $a \in \mathbb{F}_2^n$, if vectors $y_1, \dots, y_{2n}$ are chosen uniformly at random subject to $a \cdot y_j = 0$ for every j, then with high probability (i.e., probability $1 - f(n)$ where $f(n) \to 0$ as $n \to \infty$, although in this problem one can be far more precise), a subset of $n-1$ of them are linearly independent. ⊙

Thus after $2n$ runs of the algorithm, there will be $n-1$ independent hyperplanes with high probability and one finds a.

3.3. Elementary number theory needed for the algorithms of Miller-Rabin and Shor

Write $m \bmod N$ for the equivalence class of m in $\mathbb{Z}/N\mathbb{Z}$. We sometimes abuse notation and write $m \bmod N$ for its representative in $\{0, \ldots, N-1\}$.

3.3.1. The Euclidean algorithm.

Exercise 3.3.1. Let a, b be natural numbers. Show that d divides a and b if and only if d divides a and d divides $b \bmod a$.

Set $u = b \bmod a$, which we view as an element of $\{0, \ldots, a-1\}$. Exercise 3.3.1 implies

$$\gcd(a, b) = \gcd(a, u) = \gcd(a \bmod u, u), \tag{3.3.1}$$

and one can continue. This leads to the *Euclidean algorithm* for finding the greatest common divisor of two positive integers a, b. Write

$$m = qn + r,$$

where one has divided m by n and obtained q with remainder r. Assume $b < a$. One performs successive divisions:

$$\begin{aligned} a &= \alpha_1 b + r_0, \\ b &= \alpha_2 r_0 + r_1, \\ r_0 &= \alpha_3 r_1 + r_2, \\ r_1 &= \alpha_4 r_2 + r_3, \\ &\vdots \end{aligned} \tag{3.3.2}$$

This process terminates after a finite number of steps because the r_j are a strictly decreasing sequence of nonnegative integers. By continuing (3.3.1), we see the last nonzero r_m is the greatest common divisor of a and b. Also observe, by back substituting in the equations to eliminate the r_j for $j < m$, that $\gcd(a, b)$ is of the form $sa + tb$ for some integers s, t. We record this as a proposition:

Proposition 3.3.2. *For all positive integers a, b, one has $\gcd(a, b) = sa + tb$ for some integers s, t. In particular, if $\gcd(a, b) = 1$, then there exist integers s, t such that $1 = sa + tb$.*

The *Fibonacci numbers* are defined by $F_0 = 0, F_1 = 1$, and $F_n = F_{n-1} + F_{n-2}$.

The most important exercise in a first course in linear algebra (because it illustrates the power of eigenvalues and eigenvectors) shows that

$$F_n = \frac{1}{\sqrt{5}}\left[\left(\frac{1+\sqrt{5}}{2}\right)^n - \left(\frac{1-\sqrt{5}}{2}\right)^n\right],$$

which implies

$$\log(F_n) = O(n) \quad \text{and for } n > 3, \quad F_n \geq 2^{\lfloor \frac{n}{2} \rfloor}. \tag{3.3.3}$$

Exercise 3.3.3. Prove that if the Euclidean algorithm terminates after n steps, then $a \geq F_{n+2}$ and $b \geq F_{n+1}$. ⊚

Exercise 3.3.3 and Proposition 1.3.11 imply:

Proposition 3.3.4. *The Euclidean algorithm for computing* $\gcd(a,b)$ *for integers* $a > b > 0$ *may be executed with a circuit of size* $O(\log(a)^3)$.

For further discussion of implementing the Euclidean algorithm and its cost, see [**BS96**, Chap. 4].

3.3.2. Properties of $(\mathbb{Z}/N\mathbb{Z})^*$.

Exercise 3.3.5. Let $a, m \in \mathbb{Z}$ with $\gcd(a,m) = 1$. Show that for all $b \in \mathbb{Z}$ the equation $ax \equiv b \bmod m$ has a solution and that any two solutions are equivalent mod m. ⊚

Exercise 3.3.6. Let $m, n \in \mathbb{Z}$ with $\gcd(m,n) = 1$. Show that for all $b, c \in \mathbb{Z}$ the system

$$x \equiv b \bmod m$$
$$x \equiv c \bmod n$$

has a unique solution modulo mn. ⊚

Proposition 3.3.7 (Chinese Remainder Theorem). *For distinct primes p, q, there is a ring isomorphism*

$$\mathbb{Z}/pq\mathbb{Z} \cong \mathbb{Z}/p\mathbb{Z} \times \mathbb{Z}/q\mathbb{Z}.$$

More generally if $N = p_1^{a_1} \cdots p_k^{a_k}$ *with* p_j *distinct primes, there is a ring isomorphism*

$$(\mathbb{Z}/N\mathbb{Z}) \cong (\mathbb{Z}/p_1^{a_1}\mathbb{Z}) \times \cdots \times (\mathbb{Z}/p_k^{a_k}\mathbb{Z}).$$

Proof. Let $\gcd(m,n) = 1$. Consider the quotient maps $p_m : \mathbb{Z} \to \mathbb{Z}/m\mathbb{Z}$ and $p_n : \mathbb{Z} \to \mathbb{Z}/n\mathbb{Z}$ giving rise to $p_m \times p_n : \mathbb{Z} \to \mathbb{Z}/m\mathbb{Z} \times \mathbb{Z}/n\mathbb{Z}$. This is a ring map and its kernel contains $(mn)\mathbb{Z}$. Quotient the source of $p_m \times p_n$ by this kernel to obtain a ring map $\mathbb{Z}/(mn)\mathbb{Z} \to \mathbb{Z}/m\mathbb{Z} \times \mathbb{Z}/n\mathbb{Z}$. Exercise 3.3.6 implies this map is an isomorphism. Now write $N = mn$ and apply the argument iteratively until the prime factorization is obtained. □

For a ring R, let R^* denote its invertible elements under multiplication, which form a group. We also have, for $N = p_1^{a_1} \cdots p_k^{a_k}$, that

$$(\mathbb{Z}/N\mathbb{Z})^* \cong (\mathbb{Z}/p_1^{a_1}\mathbb{Z})^* \times \cdots \times (\mathbb{Z}/p_k^{a_k}\mathbb{Z})^*.$$

We will use the following classical theorem repeatedly:

Theorem 3.3.8. *If p is prime, then $(\mathbb{Z}/p\mathbb{Z})^*$ is a cyclic group of order $p-1$.*

As preparation for the proof, we introduce the following famous determinant:

Definition 3.3.9. The *Vandermonde determinant* in the variables $x_1, \dots, x_{p-1}$ is

$$\det \begin{pmatrix} 1 & 1 & \cdots & 1 & 1 \\ x_1 & x_2 & \cdots & x_{p-2} & x_{p-1} \\ x_1^2 & x_2^2 & \cdots & x_{p-2}^2 & x_{p-1}^2 \\ \vdots & \vdots & & \vdots & \vdots \\ x_1^{p-2} & x_2^{p-2} & \cdots & x_{p-2}^{p-2} & x_{p-1}^{p-2} \end{pmatrix} = \prod_{1 \le i < j \le p-1} (x_j - x_i). \tag{3.3.4}$$

For those not familiar with the Vandermonde determinant, to see that equation (3.3.4) holds at least up to scale, which is all we will need in what follows, note that both sides are homogeneous polynomials of degree $\binom{p-1}{2}$ and that their zero sets coincide. Namely the zero set of the right-hand side is the union of the $\binom{p-1}{2}$ hyperplanes $x_i = x_j$ when $i \neq j$. This set is clearly contained in the zero set of the left-hand side, but the zero set of a degree d homogeneous polynomial is a degree d hypersurface and the union of d hyperplanes has degree d, so there is nothing else in the zero set of the left-hand side. Readers unfamiliar with the geometry used in this argument may prove the equality by induction.

Proof. Let n be the maximal order of an element of $(\mathbb{Z}/p\mathbb{Z})^*$, and assume that $n \le p-2$ to obtain a contradiction.

If we set $x_j = j \in (\mathbb{Z}/p\mathbb{Z})^*$ in the determinant (3.3.4), we obtain a product of elements of $(\mathbb{Z}/p\mathbb{Z})^*$ and thus a nonzero element of $\mathbb{Z}/p\mathbb{Z}$, but our assumption is that the n-th row in (3.3.4) is $(1, \dots, 1)$, the same as the first, a contradiction. □

We found this proof in [**Con**], where 15 different proofs of the cyclicity are given. A more traditional proof utilizes the following basic facts:

(i) In an abelian group, the order of each element divides the maximal order of an element of the group.

(ii) The equation $x^d - 1 = 0$ has at most d solutions in $\mathbb{Z}/p\mathbb{Z}$.

Here is the more traditional proof: let m be the maximal order of an element. By (i), every element of $(\mathbb{Z}/p\mathbb{Z})^*$ is a solution to $x^m - 1 = 0$, which implies $m \le p-1$ and (ii) implies $p-1 \le m$.

Remark 3.3.10. In order to actually find a generator, one would need to know the prime factorization of $p-1$.

A consequence of Theorem 3.3.8 and (i) above is:

Proposition 3.3.11 (The Little Fermat Theorem). *If* $x \not\equiv 0 \bmod p$*, then*

$$x^{p-1} \equiv 1 \bmod p.$$

Exercise 3.3.12. Show that if p is prime, then for all integers m, n,

$$(m+n)^p \equiv m^p + n^p \bmod p.$$

Exercise 3.3.13. Give a different proof of the Little Fermat Theorem using Exercise 3.3.12. ⊚

We will need the following generalization of Theorem 3.3.8 for Shor's algorithm:

Theorem 3.3.14. *If p is an odd prime, then $(\mathbb{Z}/p^k\mathbb{Z})^*$ is a cyclic group of order $p^k - p^{k-1}$.*

The order is an exercise:

Exercise 3.3.15. Show that for p prime, $|(\mathbb{Z}/p^k\mathbb{Z})^*| = p^k - p^{k-1}$. ⊚

Proof of cyclicity. The proof will be by induction after we prove the case $k = 2$. We have the case $k = 1$ and now prove the case $k = 2$: let x be a generator of $(\mathbb{Z}/p\mathbb{Z})^*$.

Claim: Either x or $x + p$ is a generator of $(\mathbb{Z}/p^2\mathbb{Z})^*$.

To prove the claim, let a be the order of x in $(\mathbb{Z}/p^2\mathbb{Z})^*$, so a divides $|(\mathbb{Z}/p^2\mathbb{Z})^*| = p(p-1)$. Write $a = q(p-1) + r$ with $r < p - 1$. Then $x^a = x^{q(p-1)}x^r \equiv x^r \bmod p$, which implies $r = 0$ and $p - 1$ divides a. Thus either $a = p(p-1)$ and we are done, or $a = p - 1$. The same argument shows that the order of $x + p$ is either $p(p-1)$ or $p - 1$.

Assuming the order of x in $(\mathbb{Z}/p^2\mathbb{Z})^*$ is $p-1$, we rule out the order of $x + p$ being $p - 1$. Consider

$$\begin{aligned}(x+p)^{p-1} &\equiv x^{p-1} + (p-1)px^{p-2} \bmod p^2 \\ &\equiv 1 - px^{p-2} \bmod p^2 \\ &\not\equiv 1 \bmod p^2\end{aligned}$$

where the last line holds because $p^2 \nmid px^{p-2}$. Thus the order of $x + p$ is indeed $p(p-1)$, concluding the proof of the case $k = 2$.

The proof will be complete after we prove the following claim:

Claim: For $k \geq 2$, if z is a generator of $(\mathbb{Z}/p^k\mathbb{Z})^*$, then it is also a generator of $(\mathbb{Z}/p^{k+1}\mathbb{Z})^*$.

Arguing as above we have that the order of z in $(\mathbb{Z}/p^{k+1}\mathbb{Z})^*$ is of the form $p^u(p-1)$ some $1 \leq u \leq k$ and we want to show $u = k$. It will be sufficient to show $z^{p^{k-1}(p-1)} \not\equiv 1 \bmod p^{k+1}$.

Exercise 3.3.16. Show that for p prime and integers $b \geq 1$ and m, one has

$$(1+mp)^{p^b} \equiv 1 + mp^{b+1} \bmod p^{b+2}.$$

Since z is a generator of $(\mathbb{Z}/p^2\mathbb{Z})^*$, i.e., $(z^{p-1})^{p^1} \equiv 1 \bmod p^2$, we may write $z^{p-1} = 1 + mp$ for some m not divisible by p, as $(1+mp)^p \equiv 1 \bmod p^2$ and this fails for exponents smaller than p.

Exercise 3.3.16 implies

$$\begin{aligned} z^{p^{k-1}(p-1)} &= (z^{p-1})^{p^{k-1}} \\ &\equiv 1 + mp^k \bmod p^{k+1} \\ &\not\equiv 1 \bmod p^{k+1}. \end{aligned}$$

□

Exercise 3.3.17. Where did we use that $p \neq 2$? ⊙

3.4. Primality testing: The Miller-Rabin algorithm

The Problem. Given a natural number N, determine if N is prime.

Of course one could just test if any of the numbers $2, \ldots, \lfloor\sqrt{N}\rfloor$ divide N, but this is not efficient to compute: it is not polynomial in $\log(N)$, the number of bits of input, because there are $O(\sqrt{N})$ points to check. (If one had a list of prime numbers at most $\sqrt{N}$, one would still have $O(\frac{1}{\log(N)}\sqrt{N})$ points to check; see Theorem 3.6.3.) No efficient algorithm for testing primality was known until the 1970s when an efficient probabilistic algorithm was developed.

Recall that a probabilistic algorithm for primality testing is efficient means the following: given a natural number N as input and given access to randomness, using $\mathrm{poly}(\log(N))$ gates, the algorithm correctly outputs whether N is composite or prime with probability $> \frac{2}{3}$.

As with the problem of multiplying polynomials, where we arrived at an efficient computation working in a quotient of the polynomial ring, here we will obtain our efficient algorithm by working in a quotient of the ring of integers.

The key to the efficiency of the Miller-Rabin test is Exercise 1.3.12, where, in particular, we saw that if $a \in \{0, \ldots, N-1\}$, then $a^k \bmod N$ can be computed by a classical circuit of size $O(k\log(N)^2)$.

3.4.1. Warm-up: An efficient test to determine primality that works for "most" natural numbers. By the Little Fermat Theorem (Proposition 3.3.11), if we find x such that $x^{N-1} \not\equiv 1 \bmod N$, then we know N is composite.

Exercise 3.4.1. Show that if $\gcd(a, N) = d \neq 1$, then $a^{N-1} \not\equiv 1 \bmod N$. ⊙

Call the following probabilistic algorithm the *Fermat test*: choose $a \in \{2, \ldots, N-1\}$ uniformly at random and compute $a^{N-1} \bmod N$. If the answer is 1, the algorithm reports that N is prime and otherwise reports that it is composite. The test will always report that N is prime when it is prime, so assume N is composite. Does the algorithm correctly report that N is composite with probability greater than $\frac{2}{3}$? There are two cases:

(1) $\gcd(a, N) = d \neq 1$. (This occurs with low probability.) Then $a^{N-1} \not\equiv 1 \bmod N$ by Exercise 3.4.1 and the algorithm reports that N is composite.

(2) $\gcd(a, N) = 1$, so $a \in (\mathbb{Z}/N\mathbb{Z})^*$.

We need to determine what happens in the second case.

Introduce the following groups associated to an abelian group G: for any natural number m, consider the group homomorphism

$$\begin{aligned} \phi_m : G &\to G \\ x &\mapsto x^m. \end{aligned}$$

Let

$$G^{(m)} = \operatorname{Image} \phi_m \text{ and} \tag{3.4.1}$$

$$G_{(m)} = \ker \phi_m, \tag{3.4.2}$$

both of which are abelian groups. If a is uniformly distributed over an abelian group G, then a^m is uniformly distributed over $G^{(m)}$. This last assertion holds because for any group homomorphism of finite groups $f : G \to H$, all fibers have the same cardinality. We will use this fact repeatedly in what follows.

In this language, the a's for which the Fermat test fails are those in $(\mathbb{Z}/N\mathbb{Z})^*_{(N-1)}$.

Lemma 3.4.2. *If there exists $b \in (\mathbb{Z}/N\mathbb{Z})^*$, such that $b^{N-1} \not\equiv 1 \bmod N$, then the Fermat test detects the compositeness of N with probability $> \frac{1}{2}$.*

Proof. Assume N is composite. The hypothesis is that $(\mathbb{Z}/N\mathbb{Z})^*_{(N-1)} \neq (\mathbb{Z}/N\mathbb{Z})^*$ and we need to show at least half of the elements of $(\mathbb{Z}/N\mathbb{Z})^*$ are not in $(\mathbb{Z}/N\mathbb{Z})^*_{(N-1)}$. Since $N > 3$, the quotient $(\mathbb{Z}/N\mathbb{Z})^*/(\mathbb{Z}/N\mathbb{Z})^*_{(N-1)}$ has cardinality at least 2. Thus $a^{N-1} \not\equiv 1 \bmod N$ for at least half of the elements of $(\mathbb{Z}/N\mathbb{Z})^*$. Accounting for the divisors of N, one succeeds with probability greater than $\frac{1}{2}$. □

To succeed with probability greater than $\frac{2}{3}$, run the algorithm twice.

However, it is possible that $a^{N-1} \equiv 1 \bmod N$ for all $a \in (\mathbb{Z}/N\mathbb{Z})^*$:

Fact. The number $N = 561 = 3 * 11 * 17$ is such that $a^{N-1} \equiv 1 \bmod N$ for all $a \in (\mathbb{Z}/N\mathbb{Z})^*$. (You may check this with a computer.)

Thus we will need a second test to apply when the Fermat test reports that N is prime to obtain our desired algorithm.

3.4.2. The Miller-Rabin algorithm [Rab80]. The Fermat test exploited that if N is prime, one cannot have $a^{N-1} \not\equiv 1 \bmod N$. The *Miller-Rabin test* will exploit additional behavior that cannot happen when N is prime. Assume the Fermat test has failed for a, so $a^{N-1} - 1 \equiv 0 \bmod N$. Let 2^k be the largest power of 2 that divides $N - 1$ and write $N - 1 = 2^k\ell$. Note that $a^{2^j\ell} = (a^{2^{j-1}\ell})^2$. We have

$$\begin{aligned} a^{N-1} - 1 &= (a^{2^{k-1}\ell})^2 - 1 \\ &= (a^{2^{k-1}\ell} - 1)(a^{2^{k-1}\ell} + 1) \\ &= (a^{2^{k-2}\ell} - 1)(a^{2^{k-2}\ell} + 1)(a^{2^{k-1}\ell} + 1) \\ &\vdots \\ &= (a^{\ell} - 1)(a^{\ell} + 1)(a^{2^1\ell} + 1) \cdots (a^{2^{k-1}\ell} + 1), \end{aligned}$$

and thus

$$(a^{\ell} - 1)(a^{\ell} + 1)(a^{2^1\ell} + 1)(a^{2^2\ell} + 1) \cdots (a^{2^{k-1}\ell} + 1) \equiv 0 \bmod N.$$

Were N prime, one of these factors would have to be zero modulo N, i.e., either $a^{\ell} \equiv 1 \bmod N$ or $a^{2^j\ell} \equiv -1 \bmod N$ for some j. Therefore if this fails, we know that N is composite.

Proposition 3.4.3. *If there exists a natural number b such that $b^2 \equiv 1 \bmod N$ and $b \not\equiv \pm 1 \bmod N$, then N is composite with nontrivial factors in common with both $b + 1$ and $b - 1$.*

Proof. Under the hypotheses, $b^2 - 1 = (b - 1)(b + 1)$ is a multiple of N but $b - 1, b + 1$ are not, so N must have nontrivial factors in common with both $b + 1$ and $b - 1$. □

Remark 3.4.4. Proposition 3.4.3 may sometimes be used to efficiently find a factor of N using the Euclidean algorithm: if the order of a in $(\mathbb{Z}/N\mathbb{Z})^*$ is even, then $a \equiv c^2 \bmod N$ for some c, and we may apply Proposition 3.4.3 with $b = c^{2q}$ where $2q$ is the order of a.

Here is the Miller-Rabin algorithm: to avoid trivialities, first test if 2 divides N, so we may assume N is odd.

Choose $a \in \{2, \dots, N - 2\}$ uniformly at random.

Step 1: Test if $a^{N-1} \not\equiv 1 \bmod N$. If so, then N is composite by the Little Fermat Theorem. Otherwise go to Step 2:

Step 2: Write $N-1 = 2^k \ell$ as above. Compute the sequence

$$a^{\ell} \bmod N,\ a^{2^1\ell} \bmod N,\ a^{2^2\ell} \bmod N, \ldots, a^{2^k\ell} \bmod N.$$

If this sequence contains a 1 preceded by anything except ± 1, i.e., if there exists j such that $a^{2^j\ell} \not\equiv \pm 1 \bmod N$ and $(a^{2^j\ell})^2 \equiv 1 \bmod N$, then N is composite by Proposition 3.4.3. Otherwise the algorithm reports "N is prime".

Using Exercise 1.3.12, one sees that the total circuit size is polynomial in $\log(N)$.

Proposition 3.4.5. *The Miller-Rabin algorithm succeeds on any input with probability greater than $\frac{1}{2}$.*

Proof. It is clear that if N is prime, the algorithm always indicates that it is prime; so assume N is composite and odd. Start the algorithm with some $a \in \{2, \ldots, N-2\}$ chosen uniformly at random. If $\gcd(a, N) > 1$, then Step 1 shows that N is composite; so assume this does not happen, which implies a is uniformly distributed over $(\mathbb{Z}/N\mathbb{Z})^*$.

In order for Step 1 to work with probability at least $\frac{1}{2}$, it is enough that there is one $x \in \{2, \ldots, N-1\}$ such that $x^{N-1} \not\equiv 1 \bmod N$. The following exercise shows that if N is a power of an odd prime, then there is such an x.

Exercise 3.4.6. Show that if $N = p^c$ for some odd prime p and some $c \geq 2$, then $a = p^c + 1 - p^{c-1}$ satisfies $a^{N-1} \not\equiv 1 \bmod N$. ⊚

By Exercise 3.4.6, we may assume further that $N = uv$, where u, v are odd, $u, v > 1$, and $\gcd(u, v) = 1$. By the Chinese Remainder Theorem (Proposition 3.3.7), $(\mathbb{Z}/N\mathbb{Z})^* \cong (\mathbb{Z}/u\mathbb{Z})^* \times (\mathbb{Z}/v\mathbb{Z})^*$.

Now $(\mathbb{Z}/N\mathbb{Z})^{*(m)} \cong (\mathbb{Z}/u\mathbb{Z})^{*(m)} \times (\mathbb{Z}/v\mathbb{Z})^{*(m)}$. If either $(\mathbb{Z}/u\mathbb{Z})^{*(N-1)}$ or $(\mathbb{Z}/v\mathbb{Z})^{*(N-1)}$ is nontrivial, Step 1 will detect compositeness with probability at least $\frac{1}{2}$, so assume both are trivial. To apply Step 2, we need to consider the powers $a^{2^j\ell} \bmod N$ and show that with probability at least $\frac{1}{2}$ there exists j such that $a^{2^j\ell} \not\equiv \pm 1 \bmod N$ and $(a^{2^j\ell})^2 = a^{2^{j+1}\ell} \equiv 1 \bmod N$. Note $(\mathbb{Z}/N\mathbb{Z})^{*(\ell)} \supset \{\pm 1\}$ as $-1 \in \mathbb{Z}/N\mathbb{Z}$, and since ℓ is odd, $(-1)^{\ell} = -1$. Let j_0 be the largest value such that $(\mathbb{Z}/N\mathbb{Z})^{*(2^{j_0}\ell)} \neq \{1\}$ and $(\mathbb{Z}/N\mathbb{Z})^{*(2^{j_0+1}\ell)} = \{1\}$, so there exists $b \in (\mathbb{Z}/N\mathbb{Z})^*$ such that $b^{2^{j_0}\ell} \not\equiv 1 \bmod N$. Use the isomorphism $(\mathbb{Z}/N\mathbb{Z})^{*(2^{j_0}\ell)} \cong (\mathbb{Z}/u\mathbb{Z})^{*(2^{j_0}\ell)} \times (\mathbb{Z}/v\mathbb{Z})^{*(2^{j_0}\ell)}$: both factors cannot be trivial by assumption. If one of the two factors is trivial, we could only fail if $a^{2^{j_0}\ell}$ maps to $(1, 1)$, but this will happen for the nontrivial factor with probability at most $\frac{1}{2}$. Now assume both factors are nontrivial, say of cardinalities c_u, c_v. In this case, the image of $a^{2^{j_0}\ell}$ in the first factor is 1 with probability $\frac{1}{c_u}$, and it is 1 in the second factor with probability $\frac{1}{c_v}$, and these events are independent (again by the Chinese Remainder Theorem). Thus the probability $a^{2^{j_0}\ell} \equiv 1 \bmod N$ is $\frac{1}{c_u c_v}$. For similar reasons the probability $a^{2^{j_0}\ell} \equiv -1 \bmod N$ is either $\frac{1}{c_u c_v}$ or zero. Thus the probability of failure is at most $\frac{2}{c_u c_v} \leq \frac{1}{2}$. □

But this is not the end of the story.

3.4.3. Beyond the Miller-Rabin algorithm.

Theorem 3.4.7 ([**AKS04**]). *Primality testing is in* **P**.

The core of the proof is a variant of the Little Fermat Theorem: let a, N be relatively prime integers with $N > 2$. Then N is prime if and only if in the ring of polynomials with coefficients in $\mathbb{Z}/N\mathbb{Z}$

$$(x+a)^N \equiv x^N + a^N \bmod N. \tag{3.4.3}$$

Exercise 3.4.8. Prove N is prime if and only if (3.4.3) holds in $\mathbb{Z}/N\mathbb{Z}[x]$. ⊙

The bulk of the work is reducing the number of coefficients one needs to check in the expansion of the left-hand side.

The lesson to be drawn here is that we should not make any assumptions regarding the difficulty of a problem until we have a proof.

3.5. Quantum gate sets

So far in our quantum algorithms, we have only needed the quantum gates corresponding to classical reversible computation, the reflection $\begin{pmatrix} 1 & 0 \\ 0 & -1 \end{pmatrix}$, and the Hadamard matrix $H = \frac{1}{\sqrt{2}}\begin{pmatrix} 1 & 1 \\ 1 & -1 \end{pmatrix}$. For Shor's algorithm one needs more general gates, so in this section we discuss admissible quantum gate sets. There are two issues we need to deal with: *locality* and *approximation*. We will first see, in §3.5.1, that locality is easily dealt with. Approximation is more subtle. We will first need a way to measure how close an approximation is to a desired unitary operation. This will be done via the *operator norm* defined in §3.5.2. We introduce the "standard quantum gate set" in §3.5.3.

3.5.1. Locality. We have already discussed the need in classical computation for a gate to act on a finite number of registers, and it turned out three was sufficient. With quantum computing we further expect to only be able to implement quantum gates acting on adjacent qubits, as it is likely that one can only create entanglement in a laboratory on qubits that are physically close to one another. Thanks to the following unitary version of the classical Cartan-Dieudonné theorem, this issue is not a problem: write $|1\rangle, \dots, |n\rangle$ for the standard basis of $\mathbb{C}^n$.

Lemma 3.5.1 (Cartan-Dieudonné theorem). *Any $U \in \mathbf{U}(n)$ may be written as a product of at most $\binom{n}{2}$ elements, each of which acts on some* $\mathrm{span}\{|i\rangle, |i+1\rangle\}$ *as an element of* $\mathbf{U}(2)$ *and is the identity on the span of* $|1\rangle, \ldots, |i-1\rangle, |i+2\rangle, \ldots, |n\rangle$.

Proof. Let $\mathbf{U}(2)_i \subset \mathbf{U}(n)$ be the copy of $\mathbf{U}(2)$ acting only on $\mathrm{span}\{|i\rangle, |i+1\rangle\}$. By Exercise 3.1.1, $\mathbf{U}(2)$ acts transitively on lines in $\mathbb{C}^2$. Moreover, it can send any $|v\rangle \in \mathbb{C}^2$ to $\begin{pmatrix} |v| \\ 0 \end{pmatrix}$. Thus for any unit vector $|\psi\rangle \in \mathbb{C}^n$, there exist $U_j \in \mathbf{U}(2)_j$ such that $U_1 \cdots U_n|\psi\rangle = |1\rangle$. To find the expression for a given U, write the columns of U^{-1} as $|\psi_1\rangle, \ldots, |\psi_n\rangle$ where $|\psi_j| = 1$. We may find $U_{1,1}, \ldots, U_{1,n-1}$ such that $U_{1,1} \cdots U_{1,n-1}|\psi_1\rangle = |1\rangle$. The application of $U_{1,1}, \ldots, U_{1,n-1}$ to the remaining columns will make them orthogonal to $|1\rangle$. Next we may find $U_{2,2}, \ldots, U_{2,n-1}$ with $U_{2,j} \in \mathbf{U}(2)_j$ such that their product applied to $U_{1,1} \cdots U_{1,n-1}|\psi_2\rangle$ is $|2\rangle$. Continuing, we obtain

$$U = U_{n-1,n-1}U_{n-2,n-2}U_{n-2,n-1} \cdots U_{2,2} \cdots U_{2,n-1}U_{1,1} \cdots U_{1,n-1}. \qquad \square$$

3.5.2. Approximation. Ideally one would like to work with a universal gate set as in the classical case, but that was already not possible for probabilistic computation where we could only create probability distributions with coefficients of the form $m/2^k$, where $m, k \in \mathbb{Z}_{\geq 0}$. Similarly, when using a finite quantum gate set, we cannot generate the entire unitary group. This results in quantum computation being *approximate.* This has not been an issue so far, but it will be when we study Shor's algorithm.

We need to prove that we can get arbitrarily "close" to any $U \in \mathbf{U}(\mathcal{H})$ with a polynomial (in the input size) number of elements from our finite gate set.

First let us make the meaning of "close" precise. Instead of working with a distance function on $\mathbf{U}(\mathcal{H})$, we will work with a norm on $\mathrm{End}(\mathcal{H})$ because we will need to measure the norm of the difference of two unitary operators, which in general is not unitary. Moreover, $\mathrm{End}(\mathcal{H})$ has the advantage of being a linear space where distances are easier to work with.

Definition 3.5.2. Let V be a complex vector space. A *norm* on V is a function

$$\|\cdot\| : V \to \mathbb{R}_{\geq 0}$$

satisfying, for all $v, w \in V$ and all $c \in \mathbb{C}$,

$$\|v\| \geq 0 \text{ with equality iff } v = 0,$$
$$\|v + w\| \leq \|v\| + \|w\| \text{ (the triangle inequality)},$$
$$\|cv\| = |c|\,\|v\|.$$

In our case, $V = \mathrm{Hom}(\mathcal{H}_A, \mathcal{H}_B)$. Since $\mathrm{Hom}(\mathcal{H}_A, \mathcal{H}_B)$ inherits a Hermitian inner product from the inner products on $\mathcal{H}_A, \mathcal{H}_B$, we could just take the norm induced by the norms on $\mathcal{H}_A, \mathcal{H}_B$. However, we would like to take advantage of the extra structure of our vector space as a space of linear maps and take a norm reflecting that structure. This norm is called the *operator norm*.

Definition 3.5.3. For $X \in \text{Hom}(\mathcal{H}_A, \mathcal{H}_B)$, define the *operator norm* of X,

$$||X||_{\text{op}} := \max_{|\xi|=1} |X|\xi\rangle|$$

where $|X|\psi\rangle|$ is the usual Hermitian norm of the vector $X|\psi\rangle$.

Since the operator norm is the primary norm we will use in this chapter, we write $||X|| = ||X||_{\text{op}}$.

Exercise 3.5.4. Show that for $X \in \text{Hom}(\mathcal{H}_A, \mathcal{H}_B)$, $||X||^2$ is the largest eigenvalue of $X^\dagger X$.

Exercise 3.5.5. Show that when $X \in \text{End}(\mathcal{H})$, $||X||_{\text{op}}$ is the largest length of an eigenvalue of X.

Exercise 3.5.6. Verify that the operator norm is indeed a norm.

Remark 3.5.7. Exercise 3.5.5 shows that the operator norm is a type of ℓ_∞ norm, and it is sometimes labeled as such in the literature.

Remark 3.5.8. In the definition we have used that $\mathcal{H}_A, \mathcal{H}_B$ are finite dimensional. The operator norm is also defined on infinite-dimensional vector spaces, but because the unit sphere in infinite dimensions is not sequentially compact, one must instead define it as

$$||X||_{\text{op}} := \sup_{|\xi\rangle \neq 0} \frac{|X|\xi\rangle|}{|\xi|}.$$

Remark 3.5.9. Since we are dealing with finite-dimensional Hilbert spaces, all norms are equivalent in the sense that if a sequence goes to zero in one norm, it will in all. However, the operator norm will be particularly easy to estimate when dealing with compositions of unitary operators.

For $X \in \text{End}(\mathcal{H}_A)$, $Z \in \text{End}(\mathcal{H}_B)$, we may consider $X \otimes Z \in \text{End}(\mathcal{H}_A \otimes \mathcal{H}_B)$, where the action of $X \otimes Z$ on $v \otimes w$ is $Xv \otimes Zw$ and is extended linearly to $\mathcal{H}_A \otimes \mathcal{H}_B$. The operator norm has the following additional properties:

$$\begin{aligned} ||XY|| &\leq ||X||\,||Y||, \\ ||X^\dagger|| &= ||X||, \\ ||X \otimes Z|| &= ||X||\,||Z||, \text{ and} \\ ||U|| &= 1\ \forall U \in \mathbf{U}(\mathcal{H}). \end{aligned} \tag{3.5.1}$$

Exercise 3.5.10. Verify these additional properties.

We return to our approximation problem:

Definition 3.5.11. An operator $\tilde{U} \in \mathbf{U}(\mathcal{H})$ *approximates* $U \in \mathbf{U}(\mathcal{H})$ *with precision* δ if $\|\tilde{U} - U\| \leq \delta$.

Exercise 3.5.12. Show that $\tilde{U} \in \mathbf{U}(\mathcal{H})$ approximates $U \in \mathbf{U}(\mathcal{H})$ with precision δ if and only if $\tilde{U}^{-1}$ approximates U^{-1} with precision δ.

We will not be concerned with a single unitary transformation, but with the product of many such transformations, so we need to examine how errors grow under composition of maps. The key property of the operator norm is that for unitary transformations, errors accumulate linearly:

Proposition 3.5.13. *Say that* $U = U_L \cdots U_2 U_1$ *with* $U_j \in \mathbf{U}(\mathcal{H})$, $1 \leq j \leq L$, *and that* U_j *is approximated by* $\tilde{U}_j \in \mathbf{U}(\mathcal{H})$ *with precision* δ_j. *Then* $\tilde{U} := \tilde{U}_L \cdots \tilde{U}_2 \tilde{U}_1$ *approximates* U *with precision* $\sum_{j=1}^{L} \delta_j$.

Proof. By induction, it will be sufficient to prove the case $L = 2$. We have

$$\begin{aligned}
\|\tilde{U}_2\tilde{U}_1 - U_2U_1\| &= \|\tilde{U}_2(\tilde{U}_1 - U_1) + (\tilde{U}_2 - U_2)U_1\| \\
&\leq \|\tilde{U}_2(\tilde{U}_1 - U_1)\| + \|(\tilde{U}_2 - U_2)U_1\| \\
&\leq \|\tilde{U}_2\|\,\|\tilde{U}_1 - U_1\| + \|\tilde{U}_2 - U_2\|\,\|U_1\| \\
&\leq \|\tilde{U}_1 - U_1\| + \|\tilde{U}_2 - U_2\|.
\end{aligned}$$

□

3.5.3. The standard quantum gate set. In practice, one has to fix a quantum gate set in advance, so we will have to approximate the transformations appearing in Shor's algorithm from elements of our gate set. The following gate set has been proven to effectively approximate arbitrary unitary transformations (by Lemma 3.5.1, it is sufficient to show it can approximate elements of $\mathbf{U}(2)$ effectively).

Definition 3.5.14. The quantum gate set

$$H := \frac{1}{\sqrt{2}}\begin{pmatrix}1 & 1\\ 1 & -1\end{pmatrix} \text{ (the Hadamard gate)}, \tag{3.5.2}$$

$$K := \begin{pmatrix}1 & 0\\ 0 & i\end{pmatrix}, \tag{3.5.3}$$

$$K^{-1} = \begin{pmatrix}1 & 0\\ 0 & -i\end{pmatrix}, \tag{3.5.4}$$

$$\text{NOT} := \begin{pmatrix}0 & 1\\ 1 & 0\end{pmatrix}, \tag{3.5.5}$$

$$\text{the Toffoli gate} \tag{3.5.6}$$

is called *standard*.

In Shor's algorithm we will need to use classical bits to decide whether or not to apply a quantum gate. We now introduce notation for such classical "controls" in a quantum circuit, which will also be allowed.

For $U \in \mathbf{U}(n)$, introduce *classically controlled* U, $\Lambda^1(U) : \mathbb{C}^n \otimes \mathbb{C}^2 \to \mathbb{C}^n \otimes \mathbb{C}^2$ by

$$\Lambda^1(U)(|\xi\rangle \otimes |x\rangle) = \begin{cases} |\xi\rangle \otimes |x\rangle & \text{if } x = 0, \\ U|\xi\rangle \otimes |x\rangle & \text{if } x = 1. \end{cases}$$

When acting by these controlling bits, we will allow violation of strict locality for the controlling bits. This is physically acceptable, because it will correspond to interfering with the quantum system from "outside".

Exercise 3.5.15. Show that $\Lambda^1(\mathrm{NOT})|b, a\rangle = |a \oplus b, a\rangle$.

Theorem 3.5.16. *Any $U \in \mathbf{U}(n)$ may be realized with precision δ by a* $\mathrm{poly}(\log(\frac{1}{\delta}))$*-size circuit over the standard quantum gate set, allowing classically controlled gates and using workspace bits (i.e., working with elements of* $\mathbf{U}(n+s)$*).*

Theorem 3.5.16 justifies the assertion that one can achieve good approximate quantum algorithms from a fixed gate set. For the proof, see [**KSV02**, Thm. 13.5].

3.6. Shor's algorithm

Shor's algorithm involves a classical part and a quantum part. The quantum part is as follows: given a randomly chosen $a \in (\mathbb{Z}/N\mathbb{Z})^*$, find the order of a in $(\mathbb{Z}/N\mathbb{Z})^*$, that is, the smallest r such that $a^r \equiv 1 \bmod N$. The algorithm to do this is called the *order finding algorithm.* The classical part, which we already saw the idea of in Remark 3.4.4 and which dates back at least to 1976, uses the output of the quantum algorithm to find a factor of N. The size of the quantum circuit will be $\mathrm{poly}(\log(N))$. Since there are at most $\log(N)$ factors of N, N completely factors with $\mathrm{poly}(\log(N))$-operations.

The quantum part will hinge on (i) there being enough prime numbers less than N and (ii) the ability to closely approximate a rational number by other rational numbers. After explaining the classical part in §3.6.1, these two results are presented in §3.6.2 and §3.6.3. Next a small quantum circuit that computes the discrete Fourier transform is presented in §3.6.4. Finally the quantum part is presented in §3.6.5.

3.6.1. The classical part. Given N, the quantum part produces an a and its order r in $(\mathbb{Z}/N\mathbb{Z})^*$. We will show that with this data, with probability at least $\frac{1}{4}$, one can efficiently find a factor of N. Since the runs are independent, with four runs, we find a factor with probability greater than $\frac{2}{3}$.

When the order r is even we may write $a^r - 1 = (a^{\frac{r}{2}} - 1)(a^{\frac{r}{2}} + 1)$ and we have $(a^{\frac{r}{2}} - 1)(a^{\frac{r}{2}} + 1) \equiv 0 \bmod N$.

We already know $a^{\frac{r}{2}} \not\equiv 1 \bmod N$, so by Proposition 3.4.3, if $a^{\frac{r}{2}} \not\equiv -1 \bmod N$, we may apply the Euclidean algorithm to N and $(a^{\frac{r}{2}} - 1)$ to find a factor of N.

Thus if r is even and $a^{\frac{r}{2}} \not\equiv -1 \bmod N$, the algorithm succeeds. We will show that r is even with probability at least $\frac{1}{2}$ and if r is even, then $a^{\frac{r}{2}} \not\equiv -1 \bmod N$ with probability at least $\frac{1}{2}$ to get the asserted probability of at least $\frac{1}{4}$ of finding a factor in any given run.

Exercise 3.6.1. Let G be an abelian group of even order. Show that the number of elements of even order is at least the number of elements of odd order. ⊚

By Exercise 3.6.1, r is even with probability $\geq \frac{1}{2}$.

It remains to show that when r is even, $a^{\frac{r}{2}} \not\equiv -1 \bmod N$ with probability at least $\frac{1}{2}$.

Consider the special case $N = p_1 p_2$ where p_1, p_2 are distinct primes; $(\mathbb{Z}/N\mathbb{Z})^* \cong (\mathbb{Z}/p_1\mathbb{Z})^* \times (\mathbb{Z}/p_2\mathbb{Z})^*$. Write r_j for the order of a in $(\mathbb{Z}/p_j\mathbb{Z})^*$ and a_j for its image in $(\mathbb{Z}/p_j\mathbb{Z})^*$. We have $r = \operatorname{lcm}(r_1, r_2)$. Write $r_j = 2^{s_j} r_j'$. Without loss of generality, assume $s := s_2 \geq s_1$, so we may write $r = 2^s r'$ with r' odd.

Claim: If $s_1 < s$, then $a^{\frac{r}{2}} \not\equiv -1 \bmod N$. To see this, note that in this case r_1 divides $\frac{r}{2}$, which implies $a_1^{\frac{r}{2}} \equiv 1 \bmod p_1$. But r_2 does not divide $\frac{r}{2}$, so $a_2^{\frac{r}{2}} \not\equiv 1 \bmod p_2$. Since under the Chinese Remainder Theorem isomorphism $-1 \mapsto (-1, -1)$ we see $a^{\frac{r}{2}} \not\equiv -1 \bmod N$. (If $s_1 = s$, then one does get $a^{\frac{r}{2}} \equiv -1 \bmod N$ and the algorithm fails.)

It remains to show that $s_1 = s_2$ happens with probability at most $\frac{1}{2}$.

Let g_j be a generator of $(\mathbb{Z}/p_j\mathbb{Z})^*$. Recall from Theorem 3.3.8 that $(\mathbb{Z}/p\mathbb{Z})^*$ is a cyclic group of order $p - 1$. Our choice of random $a \in (\mathbb{Z}/N\mathbb{Z})^*$ may be considered as a choice of a random element of $[p_1 - 1] \times [p_2 - 1]$, where $(u, v) \in [p_1 - 1] \times [p_2 - 1]$ corresponds to the element $(g_1^u \bmod p_1, g_2^v \bmod p_2)$. Thus $s_1 \neq s_2$ if either u is odd and v is even or vice versa. The odds of this are $\frac{1}{4} + \frac{1}{4} = \frac{1}{2}$ and we conclude.

Exercise 3.6.2. Show that if $N = p_1^{m_1} \cdots p_h^{m_h}$ for distinct primes p_j with $h > 1$, then the same argument works to show that r is even and $a^{\frac{r}{2}} \not\equiv -1 \bmod N$ with probability at least $\frac{1}{4}$ (in fact better as long as $h > 2$).

To handle the case $N = p^m$ for some $m > 1$, one can either use a more involved argument or simply rule out this case directly as one can test if N is a power of a prime (or even the power of any integer less than $\log(N)$) using $O(\log(N)^3)$ operations. Say $k \leq \log(N) \leq k + 1$. Then (i) there are on the order of $\log(N)$ powers m to test, (ii) for each such power m we only need to check numbers between $2^{\frac{k}{m}}$ and $2^{\frac{k+1}{m}}$, and (iii) powering modulo N costs $O(m(\log(N))^3)$ by Exercise 1.3.12.

3.6.2. Preliminaries I: The number of primes. Let $\pi(x)$ denote the number of prime numbers that are less than or equal to x. The *prime number theorem*, proved independently by Hadamard and Poussin in 1896, states

$$\pi(x) \sim \frac{x}{\ln(x)};$$

more precisely,

$$\lim_{x\to\infty} \frac{\pi(x)}{x/\ln(x)} = 1. \tag{3.6.1}$$

The proof is not so simple, but for our purposes, the following result will suffice:

Theorem 3.6.3. *For all $x \geq 2$,*

$$\pi(x) \geq \frac{x}{2\log(x)}.$$

(In [**RS**$^+$**62**] it is shown that $\frac{x}{\log x} < \pi(x)$ for $x \geq 17$.) The proof is presented after several preliminary lemmas.

Notation. For a prime number p and natural number m, let $v_p(m)$ denote the largest power of p that divides m.

Thus

$$m = \prod_{p\leq m} p^{v_p(m)}.$$

Lemma 3.6.4.

$$v_p(n!) = \sum_{m\geq 1} \left\lfloor \frac{n}{p^m} \right\rfloor.$$

Proof. Among $1, 2, \ldots, n$, exactly $\lfloor \frac{n}{p} \rfloor$ are multiples of p, contributing $\lfloor \frac{n}{p} \rfloor$ to the summation, and exactly $\lfloor \frac{n}{p^2} \rfloor$ are multiples of p^2, contributing an additional $\lfloor \frac{n}{p^2} \rfloor$ to the summation (and thus multiples of p^2 are counted two times). Continuing, one obtains the result. □

We compute

$$v_p\left(\binom{2n}{n}\right) = v_p((2n)!) - 2v_p(n!) = \sum_{m\geq 1} \left\lfloor \frac{2n}{p^m} \right\rfloor - 2\left\lfloor \frac{n}{p^m} \right\rfloor.$$

Exercise 3.6.5. Show that for all $x \in \mathbb{R}$, $\lfloor 2x \rfloor - 2\lfloor x \rfloor \in \{0, 1\}$. ⊚

If $\frac{2n}{p^m} < 1$, i.e., if $m > \frac{\log(2n)}{\log(p)}$, then $\lfloor \frac{2n}{p^m} \rfloor - 2\lfloor \frac{n}{p^m} \rfloor = 0$, and thus

$$v_p\left(\binom{2n}{n}\right) \leq \frac{\log(2n)}{\log(p)}. \tag{3.6.2}$$

Stirling's formula estimates $n!$:

$$\ln(n!) = n\ln(n) - n + O(\ln(n)), \tag{3.6.3}$$

$$\log(n!) = n\log(n) - \log(e)n + O(\log(n)). \tag{3.6.4}$$

It is often proved using a contour integral of the Gamma function (see, e.g., [**Ahl78**, §5.2.5]). To see why it is plausible, write $\ln(n!) = \ln(1) + \cdots + \ln(n)$. This quantity may be estimated by

$$\int_1^n \ln(x)dx = [x\ln(x) - x]_1^n = n\ln n - n + 1,$$

giving intuition to (3.6.3).

Exercise 3.6.6. Show that $\binom{2n}{n} \geq \frac{2^{2n}}{2n}$ for n sufficiently large. (For extra credit, show it is true for all n.)

Lemma 3.6.7. $\pi(2n) \geq \frac{2n}{\log(2n)} - 1$.

Proof. By Exercise 3.6.6,

$$\begin{aligned}
2n\log(2) - \log(2n) &\leq \log\binom{2n}{n} \\
&= \log\left(\prod_{p \leq 2n} p^{v_p\left(\binom{2n}{n}\right)}\right) \\
&= \sum_{p\leq 2n} v_p\left(\binom{2n}{n}\right)\log(p) \\
&\leq \sum_{p\leq 2n} \left\lfloor \frac{\log(2n)}{\log(p)} \right\rfloor \log(p) \quad \text{by (3.6.2)} \\
&\leq \sum_{p\leq 2n} \log(2n) \\
&= \pi(2n)\log(2n).
\end{aligned}$$

Thus

$$\pi(2n) \geq \frac{2n}{\log(2n)} - 1. \tag{3.6.5}$$

□

Proof of Theorem 3.6.3. For $x \leq 16$, one can check the result by hand. Assume $x > 16$ and take n such that $16 \leq 2n \leq x < 2n + 2$. Then

$$\begin{aligned}\pi(x) &\geq \pi(2n)\\ &\geq \frac{2n}{\log(2n)} - 1\\ &= \frac{1}{2\log(2n)}[4n - 2\log(2n)]\\ &= \frac{1}{2\log(2n)}[4n - 2 - 2\log(n)]\\ &\geq \frac{1}{2\log(x)}[(x-2) + 2n - 2 - 2\log(n)]\\ &= \frac{1}{2\log(x)}[x + (2n - 4 - 2\log(n))]\\ &\geq \frac{x}{2\log(x)}.\end{aligned}$$

The third to last line holds because $2n > x - 2$ and the last line holds because $n \geq 8$. ☐

Proposition 3.6.8. *For all $r \in \mathbb{N}$, there exist at most $\log(r)$ distinct prime numbers that divide r.*

Proof. Write the decomposition of r into its prime factors as $r = p_1^{a_1} \cdots p_s^{a_s}$. Now $p_1^{a_1} \cdots p_s^{a_s} \geq 2^{a_1+\cdots+a_s} \geq 2^s$; i.e., $\log(r) \geq s$, which is what we needed to show. ☐

Proposition 3.6.9. *For $r \in \mathbb{N}$, let Σ_r denote the set of prime numbers less than $r/10$ that are relatively prime to r. Then $\#(\Sigma_r) > \frac{r}{20\log(r)} - \log(r)$.*

Proof. The natural number r has at most $\log(r)$ prime factors by Proposition 3.6.8, and by Theorem 3.6.3 there are at least $\frac{r/10}{2\log(r/10)} = \frac{r}{20(\log(r) - \log(10))} > \frac{r}{20\log(r)}$ prime numbers less than $\frac{r}{10}$. ☐

3.6.3. Preliminaries II: Continued fractions. In the Euclidean algorithm (3.3.2), α_1 is the integer part of $\frac{a}{b}$, α_2 is the integer part of $\frac{b}{r_0} = \frac{1}{(a/b)-\alpha_1}$, and for $j > 2$, α_j is the integer part of $\frac{r_{j-3}}{r_{j-2}}$.

In other words

$$\frac{a}{b} = \alpha_1 + \cfrac{1}{\alpha_2 + \cfrac{1}{\alpha_3 + \cfrac{1}{\alpha_4 + \cfrac{1}{\alpha_5 + \cdots}}}}. \tag{3.6.6}$$

Put another way, for any positive rational number the Euclidean algorithm produces a finite expression of the form appearing on the right-hand side of (3.6.6).

Given any positive real number x, we may carry out the same procedure: set $\alpha_1 = \lfloor x \rfloor$, the integer part of x; α_2, the integer part of $\frac{1}{x-\alpha_1}$; etc. This produces a sequence of integers $[\alpha_1, \alpha_2, \dots,]$. If we truncate the right-hand side of (3.6.6) when the left-hand side is x, we obtain an approximation of x.

For example, taking the expansion of π, one obtains $[3,7] = \frac{22}{7} \sim 3.142$, $[3,7,15] = \frac{333}{106} \sim 3.1415$, $[3,5,15,1] = \frac{355}{113} \sim 3.1415926$. Note how this is significantly more efficient than taking the decimal expansion.

For another example, $\frac{11}{9} = 1 + \frac{2}{9}$, so $\alpha_1 = 1$. Since $\frac{9}{2} = 4 + \frac{1}{2}$, we have $\alpha_2 = 4$, and since $2 = 2 + 0$, we have $\alpha_3 = 2$ and $\frac{11}{9} = [1,4,2]$. In the language of the Euclidean algorithm, $11 = 9 * 1 + 2, 9 = 2 * 4 + 1, 2 = 1 * 2 + 0$.

Definition 3.6.10. Call an expression of the form (3.6.6) where the α_j are integers a *continued fraction expression* for $\frac{a}{b}$ and write $\frac{a}{b} = [\alpha_1, \dots, \alpha_n]$.

As indicated by the approximations of π, continued fraction expressions provide efficient approximations.

Proposition 3.6.11. *Let $\frac{a}{b} = [\alpha_1, \dots, \alpha_n]$ be a continued fraction expression. For $1 \le k \le n$, write $[\alpha_1, \dots, \alpha_k] = \frac{p_k}{q_k}$. Set $p_0 = 1$, $p_{-1} = 0$, $q_0 = 0$, $q_{-1} = 1$. Then one has the double recursion relation*

$$\begin{aligned} p_k &= \alpha_k p_{k-1} + p_{k-2}, \\ q_k &= \alpha_k q_{k-1} + q_{k-2}. \end{aligned} \tag{3.6.7}$$

In other words

$$\begin{pmatrix} p_k & p_{k-1} \\ q_k & q_{k-1} \end{pmatrix} = \begin{pmatrix} \alpha_1 & 1 \\ 1 & 0 \end{pmatrix} \begin{pmatrix} \alpha_2 & 1 \\ 1 & 0 \end{pmatrix} \cdots \begin{pmatrix} \alpha_k & 1 \\ 1 & 0 \end{pmatrix}. \tag{3.6.8}$$

Proof. We proceed by induction. The case $k = 1$ is clear as $\alpha_1 = \frac{p_1}{q_1} = \frac{\alpha_1}{1}$. Assume that $k \ge 2$ and by induction that the result holds for $k' < k$. Equation (3.6.6) implies

$$[\alpha_1, \dots, \alpha_k] = \alpha_1 + \frac{1}{[\alpha_2, \dots, \alpha_k]}. \tag{3.6.9}$$

Write

$$\begin{pmatrix} a & b \\ c & d \end{pmatrix} = \begin{pmatrix} \alpha_2 & 1 \\ 1 & 0 \end{pmatrix} \cdots \begin{pmatrix} \alpha_k & 1 \\ 1 & 0 \end{pmatrix},$$

so $\frac{a}{c} = [\alpha_2, \ldots, \alpha_k]$ and

$$[\alpha_1, \ldots, \alpha_k] = \alpha_1 + \frac{1}{(a/c)} = \frac{a\alpha_1 + c}{a}.$$

Thus

$$\begin{pmatrix} p_k & p_{k-1} \\ q_k & q_{k-1} \end{pmatrix} = \begin{pmatrix} \alpha_1 & 1 \\ 1 & 0 \end{pmatrix} \begin{pmatrix} a & b \\ c & d \end{pmatrix}$$
$$= \begin{pmatrix} \alpha_1 & 1 \\ 1 & 0 \end{pmatrix} \begin{pmatrix} \alpha_2 & 1 \\ 1 & 0 \end{pmatrix} \cdots \begin{pmatrix} \alpha_k & 1 \\ 1 & 0 \end{pmatrix}.$$ □

Taking the determinants of both sides of (3.6.8) we obtain:

Corollary 3.6.12. $p_k q_{k-1} - p_{k-1} q_k = (-1)^k$.

Corollary 3.6.13. $\gcd(p_k, q_k) = 1$.

Exercise 3.6.14. Prove Corollary 3.6.13. ◎

Since we have an algorithm for the continued fraction decomposition based on the Euclidean algorithm, one might think that such a decomposition is unique; i.e., if $\frac{a}{b} = [\alpha_1, \ldots, \alpha_n] = [\beta_1, \ldots, \beta_m]$, then $m = n$ and $\alpha_i = \beta_i$ for all i. This is almost true:

Proposition 3.6.15. *If $\frac{a}{b} \in \mathbb{Q}$ with $\frac{a}{b} = [\alpha_1, \ldots, \alpha_n] = [\beta_1, \ldots, \beta_m]$ and $\alpha_j, \beta_j \geq 1$ for all $j > 1$, and $\alpha_n, \beta_m > 1$, then $m = n$ and $\alpha_k = \beta_k$ for all k.*

Note that if one obtains $[\alpha_1, \ldots, \alpha_n]$ via the Euclidean algorithm, then the hypotheses are satisfied.

Definition 3.6.16. Call a continued fraction expression $\frac{a}{b} = [\alpha_1, \ldots, \alpha_n]$ where $\alpha_j \geq 1$ for $j > 1$ *simple* and call a simple continued fraction *regular* if additionally $\alpha_n > 1$.

In this language, Proposition 3.6.15 says every rational number has a unique regular continued fraction decomposition.

Proof of Proposition 3.6.15. It is easy to see that $\alpha_1 = \beta_1 = \lfloor \frac{a}{b} \rfloor$. Assume by induction that $\alpha_j = \beta_j$ for all $j < k$, so

$$\frac{p_k}{q_k} = [\alpha_1, \ldots, \alpha_k] = [\alpha_1, \ldots, \alpha_{k-1}, \beta_k]$$

for some β_k. But

$$\frac{\alpha_k p_{k-1} + p_{k-2}}{\alpha_k q_{k-1} + q_{k-2}} = [\alpha_1, \dots, \alpha_k] = [\alpha_1, \dots, \alpha_{k-1}, \beta_k] = \frac{\beta_k p_{k-1} + p_{k-2}}{\beta_k q_{k-1} + q_{k-2}}$$

which implies

$$(\alpha_k - \beta_k)(p_{k-1}q_{k-2} - p_{k-2}q_{k-1}) = 0.$$

By Corollary 3.6.12, $\alpha_k = \beta_k$.

Now say $m > n$. Write $\lambda = [\beta_{n+1}, \dots, \beta_m]$. Generalizing (3.6.9), $[\beta_1, \dots, \beta_m] = [\beta_1, \dots, \beta_n] + \frac{1}{\lambda}$. But $[\beta_1, \dots, \beta_m] = [\alpha_1, \dots, \alpha_n] = [\beta_1, \dots, \beta_n]$ and we obtain a contradiction. □

However, simple continued fraction decompositions are not unique:

Proposition 3.6.17. *If $[\alpha_1, \dots, \alpha_n] = \frac{p}{q}$ is a regular continued fraction decomposition, then $[\alpha_1, \dots, \alpha_{n-1}, \alpha_n - 1, 1]$ is another simple continued fraction decomposition of $\frac{p}{q}$. As long as $\frac{p}{q}$ is in reduced form, this is the only other simple continued fraction decomposition of $\frac{p}{q}$.*

Exercise 3.6.18. Prove Proposition 3.6.17.

Proposition 3.6.19. *Let $\alpha \in \mathbb{R}_{>0}$. If the continued fraction expansion of α does not terminate at step $n-1$ with $n > 3$, then $[\alpha_1, \dots, \alpha_n] = \frac{a}{b} \in \mathbb{Q}$ with $b \geq 2^{\lfloor \frac{n}{2} \rfloor}$.*

Proof. Since $q_j \geq F_j$ we conclude by the inequality in the display (3.3.3). □

Theorem 3.6.20. *Let s, r be relatively prime integers and let $\alpha \in \mathbb{Q}$ be such that $|\frac{s}{r} - \alpha| \leq \frac{1}{2r^2}$. Then $\frac{s}{r}$ appears as some $\frac{p_k}{q_k}$ in the regular continued fraction expansion for α.*

Theorem 3.6.20 implies that if two rational numbers are close to each other, then their continued fraction expansions start out the same.

Proof. Take the regular continued fraction expansion of $\frac{s}{r} = [\beta_1, \dots, \beta_n]$, and write $\frac{p_j}{q_j} = [\beta_1, \dots, \beta_j]$.

Define λ by

$$\alpha = \frac{\lambda p_n + p_{n-1}}{\lambda q_n + q_{n-1}};$$

i.e.,

$$\lambda = \frac{p_{n-1} - \alpha q_{n-1}}{q_n \alpha - p_n}.$$

Exercise 3.6.21. Show

$$\lambda = \frac{(-1)^{n+1}}{r^2(\alpha - \frac{s}{r})} - \frac{q_{n-1}}{q_n}. \quad \circledcirc$$

Exercise 3.6.21 and the upper bound on $|\alpha - \frac{s}{r}|$ together imply $|\lambda| > 1$. By Proposition 3.6.17 we may assume $\lambda > 1$, since if the regular expansion of $\frac{s}{r}$ produces a negative λ, the other simple expansion will produce a positive λ.

Slightly abusing notation (as λ need not be an integer), write $\alpha = [\beta_1, \ldots, \beta_n, \lambda]$. Since $\lambda \in \mathbb{Q}$, we may write $\lambda = [\gamma_1, \ldots, \gamma_m]$. Note that $\gamma_1 \geq 1$ as $\lambda > 1$, so $\alpha = [\beta_1, \ldots, \beta_n, \gamma_1, \ldots, \gamma_m]$ and $\frac{s}{r}$ appears at the n-th step. □

In summary: Given $\alpha \in \mathbb{Q}$ and $N \in \mathbb{Z}$, in poly(log(N)) steps the continued fraction algorithm finds $\frac{a}{b} \in \mathbb{Q}$ such that $b \leq 16N$ and $\frac{a}{b}$ approximates α better than any other rational number with denominator at most b.

Remark 3.6.22. The statement and proof of Theorem 3.6.20 are not accurately presented in many of the standard quantum texts. For example, in [**NC00**] they simply state that one may assume $n + 1$ is even in the proof without an explanation of why.

3.6.4. Preliminaries III: A small quantum circuit that computes the discrete Fourier transform. Recall from Definition 1.2.3 the normalized DFT for $\mathbb{Z}/M\mathbb{Z}$. For $j \in \mathbb{Z}/M\mathbb{Z}$, the map is

$$|j\rangle \mapsto \frac{1}{\sqrt{M}} \sum_{k=0}^{M-1} \omega^{jk} |k\rangle$$

where $\omega = \omega_M = e^{\frac{2\pi i}{M}}$. Recall from Exercise 2.1.13 that the normalized DFT is a unitary change of basis such that in the new basis, multiplication in $\mathbb{Z}/M\mathbb{Z}$ is given by a diagonal matrix. The classical FFT writes the DFT matrix as a product of $O(\log(M))$ matrices, each with at most 2 nonzero entries in each row and column, for a total cost of $O(\log(M)M) < O(M^2)$ arithmetic operations to execute.

Write $M = 2^m$. We show that the DFT can be written as a product of $O(m^3) = O(\log(M)^3)$ controlled local unitary operators.

It will be convenient to express j in binary and view $\mathbb{C}^M = (\mathbb{C}^2)^{\otimes m}$; i.e., write

$$|j\rangle = |j_1\rangle \otimes \cdots \otimes |j_m\rangle$$

where $j = j_1 2^{m-1} + j_2 2^{m-2} + \cdots + j_m 2^0$ and $j_i \in \{0,1\}$. Recall $\omega = e^{2\pi i 2^{-m}}$. Write the DFT as

$$\begin{aligned}
&|j_1\rangle \otimes \cdots \otimes |j_m\rangle \mapsto \\
&\frac{1}{\sqrt{M}} \sum_{k_i \in \{0,1\}} \omega^{(\sum_{s=1}^m j_s 2^{m-s})(\sum_{\ell=1}^m k_\ell 2^{m-\ell})} |k_1\rangle \otimes \cdots \otimes |k_m\rangle \\
&\quad = \frac{1}{\sqrt{M}} \sum_{k_i \in \{0,1\}} \bigotimes_{\ell=1}^m \left[\omega^{(\sum_{s=1}^m j_s 2^{m-s}) k_\ell 2^{m-\ell}} |k_\ell\rangle \right] \\
&\quad = \frac{1}{\sqrt{M}} \sum_{k_i \in \{0,1\}} \bigotimes_{\ell=1}^m \left[\omega^{(j_1 2^{2m-1-\ell} + \cdots + j_m 2^{m-\ell}) k_\ell} |k_\ell\rangle \right] \\
&\quad = \frac{1}{2^{\frac{m}{2}}} (|0\rangle + \omega^{j_m 2^{m-1}} |1\rangle) \otimes (|0\rangle + \omega^{j_{m-1} 2^{m-1} + j_m 2^{m-2}} |1\rangle) \\
&\quad \otimes (|0\rangle + \omega^{j_{m-2} 2^{m-1} + j_{m-1} 2^{m-2} + j_m 2^{m-3}} |1\rangle) \\
&\quad \otimes \cdots \otimes (|0\rangle + \omega^{j_1 2^{m-1} + j_2 2^{m-2} + \cdots + j_m 2^{m-m}} |1\rangle) \qquad (3.6.10) \\
&\quad = \frac{1}{2^{\frac{m}{2}}} (|0\rangle + (e^{2\pi i})^{j_m 2^{-1}} |1\rangle) \otimes (|0\rangle + (e^{2\pi i})^{j_{m-1} 2^{-1} + j_m 2^{-2}} |1\rangle) \\
&\quad \otimes (|0\rangle + (e^{2\pi i})^{j_{m-2} 2^{-1} + j_{m-1} 2^{-2} + j_m 2^{-3}} |1\rangle) \\
&\quad \otimes \cdots \otimes (|0\rangle + (e^{2\pi i})^{j_1 2^{-1} + j_2 2^{-2} + \cdots + j_m 2^{-m}} |1\rangle), \qquad (3.6.11)
\end{aligned}$$

where in (3.6.10), if $2m - s - \ell > m$, i.e., $s + \ell < m$, there is no contribution with j_s because $\omega^{2^m} = 1$. So far we have just rewritten DFT_M in tensor product notation.

It will be notationally more convenient to write a quantum circuit for (3.6.11) with the order of factors reversed, i.e., a quantum circuit that produces

$$\begin{aligned}
&\frac{1}{\sqrt{2}} (|0\rangle + (e^{2\pi i})^{j_1 2^{-1} + j_2 2^{-2} + \cdots + j_m 2^{-m}} |1\rangle) \\
&\otimes \cdots \otimes \frac{1}{\sqrt{2}} (|0\rangle + (e^{2\pi i})^{j_{m-2} 2^{-1} + j_{m-1} 2^{-2} + j_m 2^{-3}} |1\rangle) \\
&\quad \otimes \frac{1}{\sqrt{2}} (|0\rangle + (e^{2\pi i})^{j_{m-1} 2^{-1} + j_m 2^{-2}} |1\rangle) \otimes \frac{1}{\sqrt{2}} (|0\rangle + (e^{2\pi i})^{j_m 2^{-1}} |1\rangle). \qquad (3.6.12)
\end{aligned}$$

To simplify bookkeeping, write the ambient space as $(\mathbb{C}^2)_1 \otimes \cdots \otimes (\mathbb{C}^2)_m$. Set

$$R_k = \begin{pmatrix} 1 & 0 \\ 0 & e^{\frac{2\pi i}{2^k}} \end{pmatrix}. \qquad (3.6.13)$$

Expression (3.6.12) may be obtained from $|j_1\rangle\otimes\cdots\otimes|j_m\rangle$ as follows: first apply $H = \frac{1}{\sqrt{2}}\begin{pmatrix}1 & 1\\ 1 & -1\end{pmatrix}$ to $|j_1\rangle \in (\mathbb{C}^2)_1$.

It becomes

$$\frac{1}{\sqrt{2}}(|0\rangle + e^{2\pi i j_1 2^{-1}}|1\rangle).$$

Next apply $\Lambda^1 R_2$ to $(\mathbb{C}^2)_1\otimes(\mathbb{C}^2)_2$ so the first factor becomes

$$\frac{1}{\sqrt{2}}(|0\rangle + (e^{2\pi i})^{j_1 2^{-1}+j_2 2^{-2}}|1\rangle).$$

Continue, applying $\Lambda^1 R_j$ to $(\mathbb{C}^2)_1\otimes(\mathbb{C}^2)_j$ for $j = 3,\dots,m$. At this point only the $(\mathbb{C}^2)_1$-term has been altered, as all the other factors only acted as controlling qubits.

From now on we leave the $(\mathbb{C}^2)_1$-slot alone. Next apply H to $(\mathbb{C}^2)_2$, followed by R_1. Then apply $\Lambda^1 R_{j-1}$ to $(\mathbb{C}^2)_2\otimes(\mathbb{C}^2)_j$ for $j = 3,\dots,m$.

Next apply H followed by R_2 to $(\mathbb{C}^2)_3$, then $\Lambda^1 R_{j-2}$ to $(\mathbb{C}^2)_3\otimes(\mathbb{C}^2)_j$ for $j = 4,\dots,m$. Continue until finally one just applies H to $(\mathbb{C}^2)_m$. Finally to obtain the DFT, reverse the orders of the factors (a classical operation).

In an actual quantum algorithm, we will need to approximate the R_j with elements from our gate set, so we will only obtain an approximation of the DFT.

3.6.5. The order finding algorithm. We are given a and N with $\gcd(a,N) = 1$ and we want to find the order of a in $(\mathbb{Z}/N\mathbb{Z})^*$. Set

$$m = \lceil 3\log(N)\rceil, \quad M = 2^m \sim N^3, \quad \text{and} \quad n = \lceil \log(N)\rceil.$$

Work in $(\mathbb{C}^2)^{\otimes(m+n)}$. Here and in what follows, we ignore the additional qubits needed to execute the classical reversible computations. Initialize the state to $|0^{m+n}\rangle$. Apply the Hadamard transform to each of the first m qubits to obtain

$$\frac{1}{\sqrt{M}}\sum_{x\in\mathbb{Z}/M\mathbb{Z}} |x\rangle\otimes|0^n\rangle,$$

where x is expanded out in binary. Use our (polynomial in n) reversible powering mod N routine

$$|x\rangle\otimes|y\rangle \mapsto |x\rangle\otimes|y \oplus a^x \bmod N\rangle$$

to obtain

$$\frac{1}{\sqrt{M}}\sum_{x\in\mathbb{Z}/M\mathbb{Z}} |x\rangle\otimes|a^x \bmod N\rangle.$$

Here and in what follows, $a^x \bmod N$ is to be considered as an element of $\{0,\dots, N-1\}$ expressed in binary to have $|a^x \bmod N\rangle \in (\mathbb{C}^2)^{\otimes n}$. Now perform a measurement on the last n qubits to obtain some $y_0 \in \{0,1\}^n$ that appears in the image of the map $x \mapsto a^x \bmod N$. The resulting vector is

$$\frac{1}{\sqrt{\#\{x \in \mathbb{Z}/M\mathbb{Z} \mid a^x \bmod N = y_0\}}} \sum_{x \mid a^x \bmod N = y_0} |x\rangle \otimes |y_0\rangle.$$

Let x_0 be the smallest natural number such that $a^{x_0} \equiv y_0 \bmod N$. Set r to be the unknown period and set $K = \#\{x \in \mathbb{Z}/M\mathbb{Z} \mid a^x \bmod N = y_0\}$, so K is the number of natural numbers $\ell r + x_0$ that are less than M where $\ell \geq 0$ is an integer; i.e.,

$$K = \left\lfloor \frac{M - (1 + x_0)}{r} \right\rfloor + 1. \tag{3.6.14}$$

Rewrite the output vector as

$$\frac{1}{\sqrt{K}} \sum_{\ell=0}^{K-1} |x_0 + \ell r\rangle \otimes |y_0\rangle. \tag{3.6.15}$$

Compare (3.6.15) with the step in Simon's algorithm giving rise to (3.2.2).

Apply the quantum DFT algorithm to the first m qubits to obtain, with $\omega = e^{2\pi i/M}$,

$$\frac{1}{\sqrt{M}\sqrt{K}} \sum_{x \in \mathbb{Z}/M\mathbb{Z}} \left(\sum_{\ell=0}^{K-1} \omega^{(x_0+\ell r)x} \right) |x\rangle \otimes |y_0\rangle.$$

Finally measure the first m qubits to obtain some $x \in \{0, \ldots, M-1\}$.

How will this be useful? Say the measurement produces an x such that $\frac{x}{M}$ is close to some fraction with denominator r, say $\frac{b}{r}$, more precisely that

$$\left| \frac{b}{r} - \frac{x}{M} \right| < \frac{1}{2r^2}.$$

When we take the regular continued fraction decomposition of $\frac{x}{M}$, by Theorem 3.6.20 the term $\frac{b}{r}$ will appear, so we may find r by testing the denominators q_j and taking the first one such that $a^{q_j} \equiv 1 \bmod N$.

We need to show there are enough such good x, each with sufficiently high probability of being chosen.

The probability of drawing any given $|x\rangle$ is

$$\begin{aligned} &\frac{1}{MK} |\omega^{x_0 x} + \omega^{(x_0+r)x} + \omega^{(x_0+2r)x} + \cdots + \omega^{(x_0+(K-1)r)x}|^2 \\ &= \frac{1}{MK} |1 + \omega^{rx} + \omega^{2rx} + \cdots + \omega^{(K-1)rx}|^2. \end{aligned}$$

To fix ideas, first consider the highly improbable case $M = rc$ for some natural number c. Write $\eta = \omega^{rx} = e^{2\pi i x/c}$. If $c \nmid x$, then $1+\eta+\eta^2+\cdots+\eta^{c-1} = 0$ and the probability of drawing such x is zero, but if c divides x, each power equals 1 as

$$\omega^{rjx} = \omega^{\frac{M}{c}jx} = (\omega^M)^{j\frac{x}{c}} = 1 \quad \text{because} \quad \frac{x}{c} \in \mathbb{Z},$$

and thus we will draw $x = cb$ for a random $b \in \{0, \ldots, r-1\}$ and

$$\frac{x}{M} = \frac{cb}{rc} = \frac{b}{r}.$$

All such x's are equally likely. Thus in this case one always finds r.

Now for the general case: the idea is similar — the values of x that are most likely to be drawn will be such that xr is *nearly* divisible by M.

We may repeat the experiment poly$(\log N)$ times while still being in polynomial time. Thus if G is the number of good x and P is the probability of drawing any given good x, we need $GP = \Omega(\frac{1}{\text{poly}(\log N)})$. We will show $G = \Omega(\frac{r}{\log r})$ and $P = \Omega(\frac{1}{r})$. Since $r < N$, this will suffice.

We first lower bound the number of good x: $G = \Omega(\frac{r}{\log(r)})$.

Recall that to be good the following is needed:

(1) There exists b such that $|\frac{x}{M} - \frac{b}{r}| < \frac{1}{2r^2}$.

(2) xr needs to be nearly divisible by M, e.g., $xr \bmod M < \frac{r}{10}$.

(3) b from (1) needs to be relatively prime to r.

Set $s = \lfloor \frac{xr}{M} \rfloor$, which will be our b in (1). Using s, (2) says

$$|xr - sM| < \frac{r}{10};$$

i.e.,

$$\left|\frac{x}{M} - \frac{s}{r}\right| < \frac{1}{10M}.$$

Since $M \sim N^3$ and $r < N$, we see (2) implies (1).

Lemma 3.6.23. *Fix natural numbers N and $r < N$. Write $M = 2^m$, where $m = \lceil 3\log(N) \rceil$, and $r = 2^\delta r'$ with r' odd. Consider the set of elements $x \in \mathbb{Z}/M\mathbb{Z}$ satisfying:*

(i) $0 \le xr \bmod M < \frac{r}{10}$.

(ii) $\gcd(\lfloor \frac{xr}{M} \rfloor, r) = 1$.

If $r' < 2{,}565$, there exist at least

$$\frac{r}{10{,}252\log(r)} - 5$$

such x, and otherwise there exist at least

$$r\left[\frac{1}{40\log(r')} - \frac{\log(r')}{2r'}\right]$$

such x.

Proof. Consider the case where r is odd; i.e., $\gcd(r, M) = 1$. Let $p \in \Sigma_r$, where Σ_r is the set of primes less than $r/10$ that are relatively prime to r. Take x to be the solution in $\{0, \ldots, M-1\}$ to $rx \equiv p \bmod M$ which exists and is unique by Exercise 3.3.5. Thus $p = rx - \lfloor \frac{rx}{M} \rfloor M$. Then x satisfies (i) by the definition of Σ_r. Were $\gcd(\lfloor \frac{rx}{M} \rfloor, r) > 1$, one would have $\gcd(rx - \lfloor \frac{rx}{M} \rfloor M, r) > 1$, i.e., $\gcd(p, r) > 1$, a contradiction, so (ii) is also satisfied. In this case, by Proposition 3.6.9 there are at least $\#(\Sigma_r) > \frac{r}{20\log(r)} - \log(r)$ solutions.

Next consider the special case $r = 2^\delta$ for some $\delta \le \frac{m}{3}$ (this last condition because $r < N$). Here we need $\lfloor \frac{xr}{M} \rfloor = \lfloor \frac{x}{2^{m-\delta}} \rfloor$ to be odd. We may take $x = 2^{m-\delta}(2k-1)$ for $k = 1, 2, 3, \ldots, 2^{\delta-1}$. Then $xr \equiv 0 \bmod M$ and $\frac{x}{2^{m-\delta}} = 2k - 1$ is odd. This gives at least $2^{\delta-1} = \frac{r}{2}$ values of x satisfying (i) and (ii).

The argument in the general case where r is even (and as we will see, $r' > 2{,}563$) is a combination of the above two cases. Write $r = r'2^\delta$, where r' is odd. Set $M' = \frac{M}{2^\delta} = 2^{m-\delta}$. By the case that r is odd, we have $\#(\Sigma_{r'})$ natural numbers x', $0 \le x' \le M'$, such that $x'r' \bmod M' < \frac{r'}{10}$ and $\gcd(\lfloor \frac{x'r'}{M'} \rfloor, r') = 1$.

We also have, for any $j \ge 0$,

$$1 = \gcd\left(\left\lfloor \frac{(x' + jM')r'}{M'} \right\rfloor, r'\right)$$

and if $\lfloor \frac{(x'+jM')r'}{M'} \rfloor$ is odd, we also have, since $r'/M' = r/M$,

$$1 = \gcd\left(\left\lfloor \frac{(x' + jM')r}{M} \right\rfloor, r\right).$$

We need $x' + jM' < M$, i.e., $x' + j2^{m-\delta} < 2^m$, so $j < 2^\delta - x'2^{\delta-m}$. Recalling that $x' \le M' = 2^{m-\delta}$ and that r' is odd, if $\lfloor \frac{x'r'}{M'} \rfloor$ is even, we may take any odd j with $0 \le j < 2^\delta - 1$, and if $\lfloor \frac{x'r'}{M'} \rfloor$ is odd, we may take any even j with $0 \le j < 2^\delta - 1$, so there are $2^{\delta-1}$ admissible values of j for each x'.

Finally notice that if $x_1' \neq x_2'$, we have $x_1' + j_1M' \neq x_2' + j_2M'$ as, otherwise, were $j_1 = j_2$ we would have $x_1' = x_2'$, a contradiction, and were $j_1 \neq j_2$ we would have $x_1' - x_2' = M'(j_1 - j_2)$ but the absolute value of the left-hand side is less than M' and the absolute value of the right-hand side is at least M'. We conclude in the general case where r is odd that there are at least

$$\#(\Sigma_{r'})2^{\delta-1} > \left[\frac{r'}{20\log(r')} - \log(r')\right]2^{\delta-1} = r\left[\frac{1}{40\log(r')} - \frac{\log(r')}{2r'}\right]$$

values x satisfying (i) and (ii). This takes care of all cases with $r' \geq 2{,}565$, which is the smallest r' where the term in brackets is positive.

Finally consider the case $r = 2^\delta r'$ with r' odd and bounded above by 2,563. We may take $x = 2^{m-\delta}k$ for $k \leq 2^{\delta-1}$ odd and $\gcd(k, r') = 1$. The worst case is r' is a product of four odd primes, and thus r has five prime factors, so using the same argument as in the proof of Proposition 3.6.9, we get at least

$$\begin{aligned}\frac{2^{\delta-1}}{2\log(2^{\delta-1})} - 5 &= \frac{2^{\delta-2}}{\delta-1} - 5 \\ &= \frac{r}{4r'(\log(r) - \log(r') - 1)} - 5 \\ &\geq \frac{r}{4r'(\log(r))} - 5\end{aligned}$$

good x. □

Corollary 3.6.24. *Given an increasing sequence of natural numbers N and randomly chosen $a \in (\mathbb{Z}/N\mathbb{Z})^*$ with period $r = r(a, N)$, there exist $\Omega(r/\log(r))$ elements $x \in \mathbb{Z}/M\mathbb{Z}$ satisfying:*

(i) $0 \leq xr \bmod M < \frac{r}{10}$.

(ii) $\gcd(\lfloor \frac{xr}{M} \rfloor, r) = 1$.

Proof. It is enough to observe that if N is increasing and a is chosen randomly, then on the average r will also be increasing, so one concludes by Lemma 3.6.23. □

Remark 3.6.25. The argument that there exist $\Omega(r/\log(r))$ good x is not explicitly spelled out in the literature. We thank J. M. Rojas for help in developing it.

It remains to lower bound the probability of drawing a good x: $P = \Omega(\frac{1}{r})$.

Lemma 3.6.26. *If x is such that $0 < xr \bmod M < \frac{r}{10}$, then the probability of drawing x in Shor's algorithm is $\Omega(\frac{1}{r})$.*

Proof. The probability of drawing any given x is

$$\frac{1}{KM}\left|\sum_{\ell=0}^{K-1}\omega^{\ell rx}\right|^2=\frac{1}{KM}\left|\sum_{\ell=0}^{K-1}(\omega^{rx})^\ell\right|^2. \tag{3.6.16}$$

Recalling the sum of a geometric series, this is

$$\frac{1}{KM}\left|\frac{1-\omega^{rKx}}{1-\omega^{rx}}\right|^2=\frac{1}{KM}\left|\frac{1-(\cos(\frac{2\pi rKx}{M})+i\sin(\frac{2\pi rKx}{M}))}{1-(\cos(\frac{2\pi rx}{M})+i\sin(\frac{2\pi rx}{M}))}\right|^2.$$

For small values of θ, we have the approximation $|1-\cos(\theta)-i\sin(\theta)|^2\sim\theta^2$. When x is good, $\frac{1}{M}(rx \bmod M)$ and $\frac{1}{M}(Krx \bmod M)$ are small. In this situation (3.6.16) is approximately

$$\frac{1}{KM}\left(\frac{\left[\frac{2\pi rKx \bmod M}{M}\right]^2}{\left[\frac{2\pi rx \bmod M}{M}\right]^2}\right)=\frac{1}{KM}K^2=\frac{K}{M}>\frac{1}{2r}.$$ □

3.6.6. Summary and remarks. Given relatively prime a, N, each run of Shor's algorithm finds, in polynomial time on a quantum computer, with probability $\sim\frac{1}{\text{poly}(\log(N))}$, the order of a in $(\mathbb{Z}/N\mathbb{Z})^*$. This occurs as follows: before doing anything, one checks separately that the order r is not small (e.g., less than 10,000). After an initialization step one applies the powering mod N routine to a and then measures to obtain an output vector that "sees" the order r appearing in expression (3.6.15). One then applies the quantum DFT algorithm and measures to obtain some x that with reasonably high probability has a regular continued fraction expression where r appears as one of the denominators in the sequence.

Remark 3.6.27. Given any permutation $\sigma\in\mathfrak{S}_n$ with its k-th iterate $a\mapsto\sigma^{(k)}(a)$ computable by a poly$(\log(n),\log(k))$ size circuit, the same algorithm will find the order of a under σ, i.e., the minimal r such that $\sigma^{(r)}(a)=a$.

Remark 3.6.28. Shor's algorithm and Simon's algorithm are special cases of the *hidden subgroup problem* for abelian groups: given G, a discrete group with a specific representation of its elements in binary, an explicit function $f: G\to\mathbb{F}_2^n$, and the knowledge that there exists a subgroup $G'\subset G$ such that $f(x)=f(y)$ if and only if $xy^{-1}\in G'$, find G'.

For abelian groups, it is sufficient to solve the problem for $G=\mathbb{Z}^{\oplus k}$ as all abelian groups are quotients of some $\mathbb{Z}^{\oplus k}$.

Simon's algorithm is the case $G = \mathbb{Z}_2^{\oplus m}$. The DFT_2 matrix is

$$H = \frac{1}{\sqrt{2}}\begin{pmatrix} 1 & -1 \\ -1 & 1 \end{pmatrix}$$

and G' is the subgroup generated by some fixed $a \in \mathbb{Z}_2^{\oplus m}$.

Shor's algorithm is the case $G = \mathbb{Z}$ and f is the function $x \mapsto a^x \bmod N$. Note that with Shor, as we did with multiplying polynomials, we restricted ourselves to $\mathbb{Z}/M\mathbb{Z}$ for M sufficiently large.

Both are solved via the DFT for the finite group G.

Chapter 4

Classical information theory

We have been referring to the classical unit of information as a *bit* (short for "binary digit"), an element of $\{0, 1\}$. The discovery/invention of the bit by Tukey and its development by Shannon [**Sha48**] was one of the great scientific achievements of the twentieth century, for it changed the way we view information, giving it an abstract formalism discussed in this chapter.

The basic question is: given a physical channel (e.g., telegraph wire), what is the maximal rate of transmission of information, tolerating a small amount of error?

We begin in §4.1, with toy examples leading up to the notion of channel capacity. After introducing Shannon entropy as a way to approximate multinomial coefficients, a preliminary version of Shannon's noiseless channel theorem is stated in §4.2. In §4.3, entropy is studied as a function on probability distributions with many interesting properties. In §4.4 the statement and proof of Shannon's noiseless channel theorem is given, along with several variants. In §4.5, after the introduction of conditional entropy and several toy examples, the noisy channel theorem is stated and proved.

Throughout this chapter a *channel* means a discrete, memoryless channel. Mathematically, it is described as follows:

Definition 4.0.1. Let $\mathcal{X}$, $\mathcal{Y}$ be finite sets. A (discrete) channel with input set $\mathcal{X}$ and output set $\mathcal{Y}$ is a stochastic $|\mathcal{Y}| \times |\mathcal{X}|$-matrix.

This definition goes beyond describing the physical channel. It will also be used to describe the encoding and decoding of messages.

Informally, the *capacity* of a channel is how much information may be transmitted on average with a single use of the channel. The precise definition and its justification are given in §4.5.

4.1. Data compression: Noiseless channels

4.1.1. A toy problem (following [BCHW16]). A source emits symbols from an alphabet $\mathcal{X}$ (e.g., $\{a, b, \dots, z\}$), which we want to store efficiently. Thus we try to encode the symbols to some smaller alphabet $\mathcal{Y}$, in a way that we can decode the stored data later to recover the original symbols.

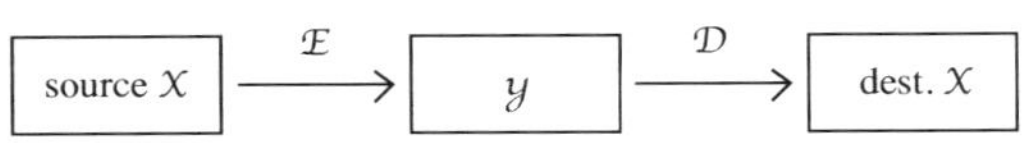

Figure 4.1.1. Message from source encoded into smaller alphabet and then decoded

In the language of channels, we seek an encoding channel $\mathcal{E} : \mathbb{R}^{|\mathcal{X}|} \to \mathbb{R}^{|\mathcal{Y}|}$ and a decoding channel $\mathcal{D} : \mathbb{R}^{|\mathcal{Y}|} \to \mathbb{R}^{|\mathcal{X}|}$. Since $\mathbb{R}^{|\mathcal{X}|}$ has a basis indexed by $\mathcal{X}$ and similarly for $\mathbb{R}^{|\mathcal{Y}|}$ and since the encoder and decoder will take basis vectors to basis vectors, we sometimes slightly abuse notation and write the maps as $\mathcal{E} : \mathcal{X} \to \mathcal{Y}$ and $\mathcal{D} : \mathcal{Y} \to \mathcal{X}$.

The symbols from $\mathcal{X}$ may be emitted with varying frequencies, but they are assumed to follow some known probability distribution on $\mathcal{X}$. Let $p = P_{\mathcal{X}}$ denote the associated probability distribution. For example, if $\mathcal{X}$ consists of the letters $\{a, \dots, z\}$ and the symbols sent correspond to messages in English, then a and e would appear more frequently than q and x. We want to determine the minimum possible size for $\mathcal{Y}$. Since we will soon be dealing with bits, it will be convenient to use the logarithms of cardinalities. Define

$$\mathrm{Cap}^0(P_{\mathcal{X}}) := \min \log |\mathcal{Y}|.$$

(Recall the notation $\log = \log_2$.)

Consider the case $\mathcal{X} = \{a, b, c, d\}$ where $p(a) = 0.1$, $p(b) = 0$, $p(c) = 0.4$, and $p(d) = 0.5$. We can clearly get away with $\mathcal{Y}$ having cardinality 3; e.g., for the encoding channel, send a, b to 1, c to 2, and d to 3; then for the decoding channel, send 1 to a, 2 to c, and 3 to d. We can always throw away symbols with probability zero. On the other hand, we cannot map two distinct symbols that do occur to the same symbol, as there would be no way to distinguish them when decoding. Thus $\mathrm{Cap}^0(p) = \log \mathrm{supp}(p)$, where

$$\mathrm{supp}(p) = \#\{x \in \mathcal{X} \mid p(x) > 0\}.$$

Now say we are willing to tolerate a small error $\epsilon > 0$. In addition to throwing out the symbols that do not appear, we may also discard the largest set of symbols whose total probability is smaller than ϵ. Let $\mathrm{Cap}^{\epsilon}(p)$ denote the corresponding minimal $\log |\mathcal{Y}|$. In the example above, if we take $\epsilon > 0.1$, we can lower storage cost, taking $|\mathcal{Y}| = 2$.

Recall that a probability distribution p on $\mathcal{X}$ satisfies $\sum_{x\in\mathcal{X}} p(x) = 1$. Relax this to *nonnormalized* probability distributions:

$$q : \mathcal{X} \to [0,1], \quad \text{satisfying} \quad \sum_{x\in\mathcal{X}} q(x) \leq 1.$$

Define

$$\mathrm{Cap}^{\epsilon}(p) := \min \log \mathrm{supp}(q),$$

where the min is taken over all nonnormalized probability distributions q satisfying $q(x) \leq p(x)$ for all $x \in \mathcal{X}$ and $\sum_{x\in\mathcal{X}} q(x) \geq 1 - \epsilon$.

4.1.2. The case of interest. Now say the transmission is not a single symbol, but a string of d symbols. We seek an encoding channel $\mathcal{E}^d : \mathcal{X}^d \to \mathcal{Y}(d)$, where $\mathcal{X}^d = \mathcal{X}\times\cdots\times\mathcal{X}$ is d copies of $\mathcal{X}$ with the induced probability distribution p^d, and decoding channel $\mathcal{D}^d : \mathcal{Y}(d) \to \mathcal{X}^d$. We want to minimize $|\mathcal{Y}(d)|$, with a tolerance of error that goes to zero as d goes to infinity. Let $\Lambda_C^d := \mathcal{D}^d \circ \mathcal{E}^d$ denote the channel obtained by composition. (The subscript C, for "classical", is used to distinguish classical channels from quantum channels which will generally be denoted Λ.) We formalize this notion of tolerable error:

Definition 4.1.1. A sequence of channels $\Lambda_C^d : \mathbb{R}^{|\mathcal{X}|^d} \to \mathbb{R}^{|\mathcal{X}|^d}$ *reliably transmits* information if, as $d \to \infty$, the probability of error goes to zero.

Remark 4.1.2. In practice one wants to send information through a communication channel. The channel can only send a limited number of bits per second, and we want to maximize the amount of information we can reliably send per second. As the length of time increases, the length of the message one can send increases.

Definition 4.1.3. Let $\mathcal{X}$ be equipped with a probability distribution p. Define the *capacity*

$$\mathrm{Cap}(p) := \lim_{\epsilon\to 0} \lim_{d\to\infty} \frac{1}{d} \mathrm{Cap}^{\epsilon}(p^d). \tag{4.1.1}$$

It remains to compute the right-hand side of (4.1.1).

We will have more general notions of capacity. In the notation of the more general notion of capacity of Definition 4.5.2 below, $\mathrm{Cap}(p) = \mathrm{Cap}(\mathrm{Id}_{\mathcal{H}}, \bar{p})$.

4.2. Preliminary solution to the compression problem

4.2.1. Estimating $|\mathcal{Y}(d)|$. At this point, readers unfamiliar with the weak law of large numbers should read §B.3.

Let $\mathcal{X} = \{1, \ldots, n\}$, and let $\mathcal{X}$ be equipped with a probability distribution $p_{\mathcal{X}}$. Introduce the notation $\bar{p} = (p_1, ..., p_n)$ to encode the probabilty distribution as a vector. Write $\bar{x}^d = (x_1, \ldots, x_d)$ for an element of $\mathcal{X}^d$. Define a map

$$\begin{aligned} \mathrm{wt}_d \, : \, & \mathcal{X}^d \to \mathbb{R}^n \\ & \bar{x}^d \mapsto \bar{c} := (c_1, \ldots, c_n), \end{aligned}$$

where c_j is the number of times j occurs in the string $(x_1, \ldots, x_d)$, so wt_d is actually $\mathbb{Z}^n_{\geq 0}$-valued. View wt_d as an n-tuple of random variables or, better, as an $\mathbb{R}^n$-valued random variable. The expectation of wt_d is

$$E[\mathrm{wt}_d] = (dp_1, \ldots, dp_n).$$

Since the string $(x_1, \ldots, x_d)$ is iid, the weak law of large numbers (B.3.1) implies that for any $\epsilon > 0$,

$$\lim_{d\to\infty} \Pr\left[\left\| \frac{1}{d}(\mathrm{wt}_d - E[\mathrm{wt}_d]) \right\|_1 \geq \epsilon \right] = 0,$$

where for $f : \mathcal{Z} \to \mathbb{R}^n$, define $\|f\|_1 = \sum_{z\in\mathcal{Z}} \|f(z)\|_1$. In our case $\mathcal{Z} = \mathcal{X}^d$.

To minimize $|\mathcal{Y}(d)|$ while having error going to zero as $d \to \infty$, we throw out all strings $\bar{x}^d$ with $\|\frac{1}{d}(\mathrm{wt}_d(\bar{x}^d) - E[\mathrm{wt}_d])\|_1 \geq \epsilon$; i.e., take

$$\mathcal{Y}(d) = \left\{ \bar{x}^d \,\middle|\, \left\| \frac{1}{d}(\mathrm{wt}_d(\bar{x}^d) - E[\mathrm{wt}_d]) \right\|_1 < \epsilon \right\}, \tag{4.2.1}$$

so $\mathcal{Y}(d)$ will have size

$$\begin{aligned} |\mathcal{Y}(d)| &= \# \left\{ \bar{x}^d \,\middle|\, \left\| \frac{1}{d}(\mathrm{wt}_d(\bar{x}^d) - E[\mathrm{wt}_d]) \right\|_1 < \epsilon \right\} \\ &= \sum_{\substack{\text{weights } \bar{c} | \\ \sum c_j = d \text{ and } \|\frac{1}{d}(\bar{c} - E[\mathrm{wt}_d])\|_1 < \epsilon}} \# \left\{ \bar{x}^d \mid \mathrm{wt}_d(\bar{x}^d) = \bar{c} \right\} \\ &= \sum_{\substack{\text{weights } \bar{c} | \\ \sum c_j = d \text{ and } \|\frac{1}{d}(\bar{c} - E[\mathrm{wt}_d])\|_1 < \epsilon}} \binom{d}{\bar{c}}. \end{aligned}$$

Here the multinomial coefficient

$$\binom{d}{\bar{c}} := \frac{d!}{c_1! \cdots c_n!} \tag{4.2.2}$$

is the number of $\bar{x}^d$ with $\mathrm{wt}_d(\bar{x}^d) = \bar{c}$.

If ϵ is small, all the $\bar{c}$ will be very close to the expectation $(dp_1, \dots, dp_n)$, and we may closely approximate all the multinomial coefficients with

$$\binom{d}{dp_1, \dots, dp_n} = \frac{d!}{(p_1 d)! \cdots (p_n d)!}. \tag{4.2.3}$$

We are cheating a little here as dp_j might not be an integer. We will be more precise later. Later we will also prove that the number of $\bar{c}$ close to the expectation has growth bounded by a polynomial in d, so we may take

$$|\mathcal{Y}(d)| \le \mathrm{poly}(d) \binom{d}{dp_1, \dots, dp_n}. \tag{4.2.4}$$

We next approximate (4.2.3).

4.2.2. Approximating binomial coefficients. Recall Stirling's formula (3.6.4):

$$\log(d!) = d\log(d) - \log(e)d + O(\log(d)).$$

In particular, for $0 < \beta < 1$ such that $\beta d \in \mathbb{Z}$,

$$\begin{aligned} \log \binom{d}{\beta d} &= \log \frac{d!}{(\beta d)!\,((1-\beta)d)!} \\ &= d[-\beta \log(\beta) - (1-\beta)\log(1-\beta)] + O(\log(d)). \end{aligned} \tag{4.2.5}$$

Let

$$H(\beta) := -\beta \log(\beta) - (1-\beta)\log(1-\beta),$$

and more generally:

Definition 4.2.1. For a probability distribution $\bar{p} = (p_1, \dots, p_n)$ on $\mathcal{X} = \{1, \dots, n\}$, the *Shannon entropy* of $\bar{p}$ is

$$H(\bar{p}) = -\sum_{i=1}^{n} p_i \log(p_i).$$

Here and throughout, we adopt the convention that $0\log(0) := 0$ as

$$\lim_{k\to\infty} 2^{-k}\log(2^{-k}) = \lim_{k\to\infty} -k2^{-k} = 0.$$

Shannon entropy plays a central role in information theory.

Exercise 4.2.2. Show that for the multinomial coefficient

$$\binom{d}{p_1 d, \dots, p_n d},$$

where $p_1 + \cdots + p_n = 1$ and $p_j d \in \mathbb{N}$, we have

$$\log \binom{d}{p_1 d, \dots, p_n d} = dH(\bar{p}) + O(\log(d)). \tag{4.2.6}$$

4.2.3. Shannon's noiseless channel theorem, preliminary version. When we take logarithms, the right-hand side of (4.2.4) becomes $dH(\bar{p}) + O(\log(d))$. Thus

$$\frac{1}{d}\log|\mathcal{Y}(d)| \leq H(\bar{p}) + o(1),$$

and we see $\mathrm{Cap}(\bar{p}) \leq H(\bar{p})$.

Theorem 4.2.3 ([**Sha48**]). $\mathrm{Cap}(\bar{p}) = H(\bar{p})$.

We give a proof in §4.4. Since the entropy $H(\bar{p})$ is a fundamental quantity, we first discuss some of its properties.

4.3. Entropy, i.e., uncertainty

Entropy is a measure of the uncertainty of an outcome. For example, consider the case $n = 2$, with probabilities $p, 1-p$. The graph of $H(p)$ is given in Figure 4.3.1.

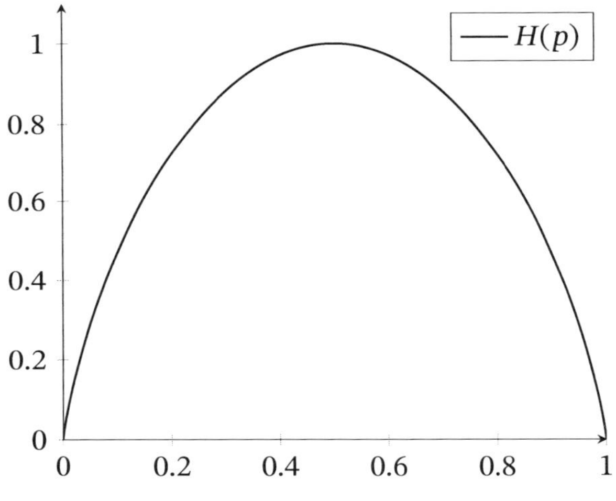

Figure 4.3.1. Graph of $H(p)$

The function $H(p)$ has a unique maximum value of 1 when $p = \frac{1}{2}$ (situation of maximal uncertainty) and it is 0 in the two certain cases.

We will be discussing properties of entropy that are consequences of concavity. We take a brief mathematical detour.

4.3.1. Concavity.

Definition 4.3.1. Let $I \subseteq \mathbb{R}$ be an interval. A function $f : I \to \mathbb{R}$ is said to be *concave* if the line segment joining any two points of the graph of f lies underneath the graph of f; i.e., for all $s \in [0,1]$ and $x_1, x_2 \in I$,

$$sf(x_1) + (1-s)f(x_2) \leq f(sx_1 + (1-s)x_2).$$

For example, if f is differentiable and for all $x \in I$, $f''(x) < 0$, then f is concave. In particular, for $x > 0$, $\log(x)$ is concave.

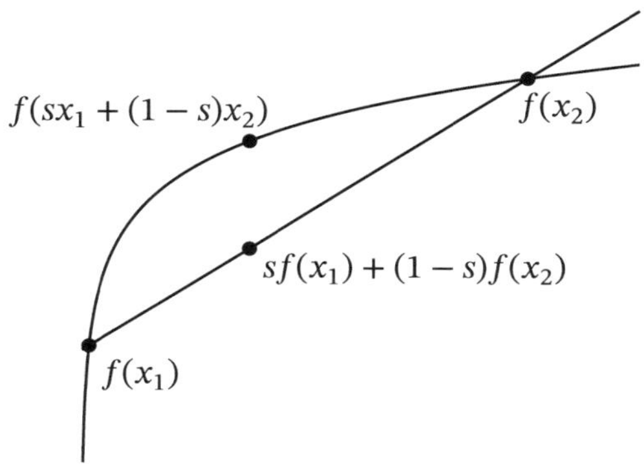

Figure 4.3.2. A concave function

More generally, if $\bar{p}$ is a probability distribution and f is concave, a straightforward induction argument gives

$$\sum p_i f(x_i) \leq f\left(\sum p_i x_i\right), \tag{4.3.1}$$

which is called *Jensen's inequality*.

In particular, if f is concave and X is a random variable, one has

$$f(E[X]) \geq E(f(X)).$$

If f is *convex*, i.e., the graph of line segments lies above the graph of f, one has the corresponding inequalities in the other direction.

4.3.2. A characterization of Shannon entropy. Shannon entropy H satisfies the following properties:

(1) $H(\frac{1}{2}, \frac{1}{2}) = 1$.
(2) $H(p_1, \dots, p_n)$ is continuous in the p_i.
(3) If $p_i = \frac{1}{n}$ for all i and we increase n, H increases.
(4) H is compatible with the law of total probability: write $\mathcal{X} := \{1, ..., n\} = A_1 \sqcup \cdots \sqcup A_k$ and assume $\Pr(A_\ell) > 0$ for all ℓ. Write $\bar{p}_A = (\Pr(A_1), \dots, \Pr(A_k))$. Then

$$H(\bar{p}_{\mathcal{X}}) = H(\bar{p}_A) + \sum_{\ell=1}^{k} \Pr(A_\ell) H(\bar{p}_{\mathcal{X}|A_\ell}).$$

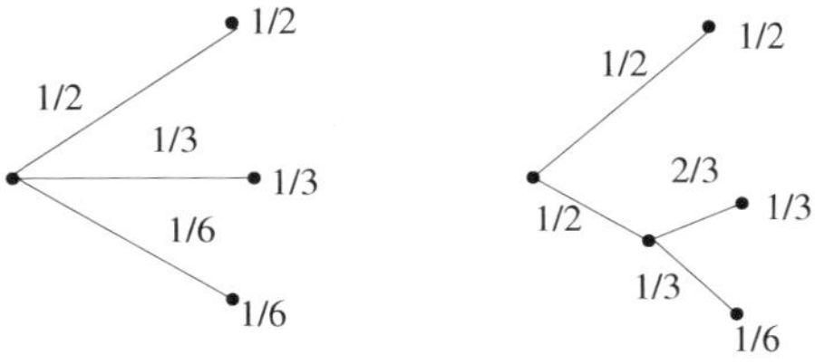

Figure 4.3.3. A choice of three outcomes viewed as a choice of two followed by a second choice of two

Here is an illustration of (4): if $\mathcal{X} = \{1, 2, 3\}$, with $p_1 = \frac{1}{2}$, $p_2 = \frac{1}{3}$, $p_3 = \frac{1}{6}$, we can compute $H(\frac{1}{2}, \frac{1}{3}, \frac{1}{6})$ by first deciding between the sets $A_1 = \{1\}$ and $A_2 = \{2, 3\}$, both of which have probability $\frac{1}{2}$. If we choose the second, we then decide between 2 and 3, the first with probability $\frac{2}{3}$, the second with probability $\frac{1}{3}$. See Figure 4.3.3. That is,

$$H\left(\frac{1}{2}, \frac{1}{3}, \frac{1}{6}\right) = H(\bar{p}_A) + \Pr(A_1)H(\bar{p}_{\mathcal{X}|A_1}) + \Pr(A_2)H(\bar{p}_{\mathcal{X}|A_2})$$
$$= H\left(\frac{1}{2}, \frac{1}{2}\right) + \frac{1}{2}(0) + \frac{1}{2}H\left(\frac{2}{3}, \frac{1}{3}\right).$$

We arrived at Shannon entropy while approximating logs of multinomial coefficients. Say we had not yet discovered it but we were looking for a function h on probability distributions with the properties (1)–(4).

Theorem 4.3.2 (**[Sha48]**). *Let h be a function satisfying* (1)–(4). *Then $h = H$ is the entropy.*

Proof. Set $A(n) = h(\frac{1}{n}, \dots, \frac{1}{n})$. Claim: By (4), $A(s^m) = mA(s)$. To see this, consider the example $s = 2, m = 3$ in Figure 4.3.4.

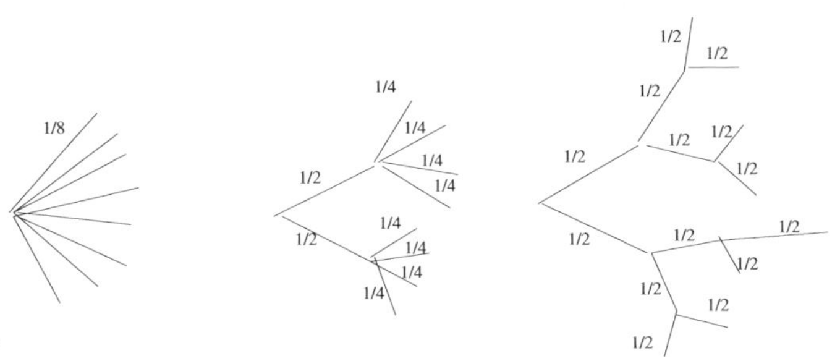

Figure 4.3.4. Repeated application of (4) gives $A(2^3) = h(\frac{1}{8}, \dots, \frac{1}{8}) = h(\frac{1}{2}, \frac{1}{2}) + [\frac{1}{2}h(\frac{1}{4}, \dots, \frac{1}{4}) + \frac{1}{2}h(\frac{1}{4}, \dots, \frac{1}{4})]$ and $h(\frac{1}{4}, \dots, \frac{1}{4}) = h(\frac{1}{2}, \frac{1}{2}) + [\frac{1}{2}h(\frac{1}{2}, \frac{1}{2}) + \frac{1}{2}h(\frac{1}{2}, \frac{1}{2})]$.

Exercise 4.3.3. Use the claim and properties (1) and (3) to show that for all $s \in \mathbb{N}, A(s) = -\log(\frac{1}{s})$. ⊚

So far we have that h agrees with H on uniform distributions. To extend to arbitrary rational distributions we use (4) again:

Say we have a choice of D equally likely outcomes, which we break up as $D = \sum_{j=1}^{n} d_j$ with $d_j \in \mathbb{N}$. Write $p_j = \frac{d_j}{D}$.

By property (4),

$$\begin{aligned} -\log\left(\frac{1}{D}\right) &= h\left(\frac{1}{D}, \ldots, \frac{1}{D}\right) \\ &= h(p_1, \ldots, p_n) + \sum_j p_j h\left(\frac{1}{d_j}, \ldots, \frac{1}{d_j}\right) \\ &= h(p_1, \ldots, p_n) - \sum_j p_j \log\left(\frac{1}{d_j}\right); \end{aligned}$$

i.e.,

$$\begin{aligned} h(p_1, \ldots, p_n) &= \sum_j p_j \log\left(\frac{1}{d_j}\right) - \log\left(\frac{1}{D}\right) \\ &= \sum_j p_j \log\left(\frac{1}{d_j}\right) - \sum_j p_j \log\left(\frac{1}{D}\right) \\ &= -\sum_j p_j \log\left(\frac{d_j}{D}\right) \\ &= -\sum_j p_j \log(p_j). \end{aligned}$$

Finally, use property (2) to extend by continuity to all probability distributions. □

Recall doubly stochastic matrices from from §1.6. Doubly stochastic matrices tend to equalize probability distributions; e.g.,

$$\begin{pmatrix} \frac{1}{2} & \frac{1}{2} \\ \frac{1}{2} & \frac{1}{2} \end{pmatrix} \begin{pmatrix} p \\ 1-p \end{pmatrix} = \begin{pmatrix} \frac{1}{2} \\ \frac{1}{2} \end{pmatrix}.$$

Proposition 4.3.4. *Entropy satisfies the following additional properties:*

(1) $0 \leq H(\bar{p}) \leq \log(n)$. *Moreover* $H(\bar{p}) = 0$ *if and only if the outcome is certain, i.e.,* $\bar{p} = (0, \ldots, 0, 1, 0, \ldots, 0)$, *and* $H(\bar{p}) = \log(n)$ *if and only if* $\bar{p} = (\frac{1}{n}, \ldots, \frac{1}{n})$.

(2) H *is subadditive: say our space is* $\mathcal{X} \times \mathcal{Y}$, *with probability distribution* $p_{\mathcal{X}\times\mathcal{Y}}$ *and marginal distributions* $p_{\mathcal{X}}$, $p_{\mathcal{Y}}$ *(cf. Definition* B.1.2*). Then*

$$H(\bar{p}_{\mathcal{X}\times\mathcal{Y}}) \leq H(\bar{p}_{\mathcal{X}}) + H(\bar{p}_{\mathcal{Y}}), \tag{4.3.2}$$

with equality if and only if the marginal distributions $p_{\mathcal{X}}$ *and* $p_{\mathcal{Y}}$ *are independent. In words: the uncertainty of* $p_{\mathcal{X}\times\mathcal{Y}}$ *is at most the sum of the uncertainties of the marginal distributions* $p_{\mathcal{X}}$ *and* $p_{\mathcal{Y}}$.

(3) *A modification of the* p_j *towards equalization increases* H. *More precisely, let* A *be a doubly stochastic matrix. Then* $H(A\bar{p}) \geq H(\bar{p})$ *and unless* A *is a permutation matrix, there exists* $\bar{p}$ *where the inequality is strict.*

(4) *Entropy is a concave function of its arguments; i.e., for all* $0 \leq t \leq 1$,

$$H(t\bar{q} + (1-t)\bar{p}) \geq tH(\bar{q}) + (1-t)H(\bar{p}).$$

Proof. (1) is safely left to the reader.

To see (2):

$$\begin{aligned}
&H(\bar{p}_{\mathcal{X}\times\mathcal{Y}}) - H(\bar{p}_{\mathcal{X}}) - H(\bar{p}_{\mathcal{Y}}) \\
&\quad= -\sum_{x\in\mathcal{X}, y\in\mathcal{Y}} p_{\mathcal{X}\times\mathcal{Y}}(x,y)\log(p_{\mathcal{X}\times\mathcal{Y}}(x,y)) \\
&\qquad+ \sum_{x\in\mathcal{X}, y\in\mathcal{Y}} p_{\mathcal{X}\times\mathcal{Y}}(x,y)\log(p_{\mathcal{X}}(x)) + \sum_{x\in\mathcal{X}, y\in\mathcal{Y}} p_{\mathcal{X}\times\mathcal{Y}}(x,y)\log(p_{\mathcal{Y}}(y)) \\
&\quad= \sum_{x\in\mathcal{X}, y\in\mathcal{Y}} p_{\mathcal{X}\times\mathcal{Y}}(x,y)\log\left(\frac{p_{\mathcal{X}}(x)p_{\mathcal{Y}}(y)}{p_{\mathcal{X}\times\mathcal{Y}}(x,y)}\right) \\
&\quad\leq \log\left(\sum_{x\in\mathcal{X}, y\in\mathcal{Y}} p_{\mathcal{X}\times\mathcal{Y}}(x,y)\frac{p_{\mathcal{X}}(x)p_{\mathcal{Y}}(y)}{p_{\mathcal{X}\times\mathcal{Y}}(x,y)}\right) \\
&\quad= \log\left(\left(\sum_{x\in\mathcal{X}} p_{\mathcal{X}}(x)\right)\left(\sum_{y\in\mathcal{Y}} p_{\mathcal{Y}}(y)\right)\right) \\
&\quad= \log((1)(1)) \\
&\quad= 0,
\end{aligned}$$

where for the inequality we used Jensen's inequality.

Exercise 4.3.5. Prove (3).

Exercise 4.3.6. Prove (4). □

We give a quantitative version of (3) in §6.5.1.

Remark 4.3.7. Property (3) says that channels which correspond to doubly stochastic matrices do not decrease entropy. This is not true for all channels. Consider

$$\Lambda_C = \begin{pmatrix} 1 & \cdots & 1 \\ 0 & \cdots & 0 \\ \vdots & \vdots & \vdots \\ 0 & \cdots & 0 \end{pmatrix}.$$

4.4. Shannon's noiseless channel theorem

Theorem 4.4.6 will be a precise version of Theorem 4.2.3. We begin with a discussion of the compression space (4.2.1).

4.4.1. Compression spaces and preliminary results.

Theorem 4.4.1 (Typical subspace theorem, [**Sha48**]). *Given $\mathcal{X} = \{a_1, \dots, a_n\}$ with probability distribution $\bar{p}$ and $\epsilon, \delta > 0$, there exists d_0 such that for all $d \geq d_0$, there is a decomposition $\mathcal{X}^d = \mathcal{X}^d_{\epsilon-\text{typ}} \sqcup \mathcal{X}^d_{\delta-\text{small}}$ where*

(1)
$$\Pr(\mathcal{X}^d_{\delta-\text{small}}) < \delta$$

and

(2)
$$(1-\delta)2^{d(H(\bar{p})-\epsilon)} \leq |\mathcal{X}^d_{\epsilon-\text{typ}}| \leq 2^{d(H(\bar{p})+\epsilon)}.$$

To interpret (2),

$$2^{dH(\bar{p})} \sim \binom{d}{dp_1, \dots, dp_n},$$

making the connection with (4.2.4), and

$$|\mathcal{X}^d| = n^d = 2^{d\log(n)},$$

so if $\bar{p}$ is far from uniform, we can have high compression, and if it is close to uniform, we get almost none.

Proof. Take

$$\mathcal{X}^d_{\epsilon-\text{typ}} = \mathcal{Y}(d) = \left\{ \bar{x}^d \;\middle|\; \left\| \frac{1}{d}(\text{wt}_d(\bar{x}^d) - E[\text{wt}_d]) \right\|_1 < \frac{\epsilon}{\log(n)} \right\},$$

and take $\mathcal{X}^d_{\delta-\text{small}}$ its complement. Recall that $E[\text{wt}_d] = (dp_1, \dots, dp_n)$. We immediately obtain (1) from the law of large numbers (B.3.2).

To prove (2), we use the following inequalities:

(4.4.1) $1 - \delta \leq \Pr(\mathcal{X}^d_{\epsilon-\mathrm{typ}}) \leq 1,$

and

(4.4.2) $|\mathcal{X}^d_{\epsilon-\mathrm{typ}}| \min_{\bar{x}\in\mathcal{X}^d_{\epsilon-\mathrm{typ}}} \bar{p}(\bar{x}) \leq \Pr(\mathcal{X}^d_{\epsilon-\mathrm{typ}}) \leq |\mathcal{X}^d_{\epsilon-\mathrm{typ}}| \max_{\bar{x}\in\mathcal{X}^d_{\epsilon-\mathrm{typ}}} \bar{p}(\bar{x}).$

Given $\bar{x} \in \mathcal{X}^d_{\epsilon-\mathrm{typ}}$ write $\mathrm{wt}_d(\bar{x}) = (c_1, \dots, c_n)$, and recall that $\sum c_j = d$. By the definition of $\mathcal{X}^d_{\epsilon-\mathrm{typ}}$ we have $\sum_{j=1}^n |c_j - dp_j| < d\frac{\epsilon}{\log(n)}$. Let $d\epsilon_j := c_j - dp_j$, so $\sum \epsilon_j = 0$ because the exponents must sum to d, and $\sum |\epsilon_j| < \frac{\epsilon}{\log(n)}$. Then

$$\begin{aligned} p(\bar{x}) &= p_1^{c_1} \cdots p_n^{c_n} \\ &= p_1^{d(p_1+\epsilon_1)} \cdots p_n^{d(p_n+\epsilon_n)} \\ &= p_1^{dp_1} \cdots p_n^{dp_n} p_1^{d\epsilon_1} \cdots p_n^{d\epsilon_n} \\ &= 2^{-dH(\bar{p})} 2^{d(\epsilon_1 p_1 \log(p_1) + \cdots + \epsilon_n p_n \log(p_n))}. \end{aligned}$$

Note that $-\epsilon \leq \epsilon_1 p_1 \log(p_1) + \cdots + \epsilon_n p_n \log(p_n) \leq \epsilon$; i.e., for all $\bar{x} \in \mathcal{X}^d_{\epsilon-\mathrm{typ}}$,

(4.4.3) $$2^{-d(H(\bar{p})+\epsilon)} \leq p(\bar{x}) \leq 2^{-d(H(\bar{p})-\epsilon)}.$$

Combining this estimate with (4.4.1), (4.4.2) proves (2). □

It will be useful to give two variants of Theorem 4.4.1:

Theorem 4.4.2 ([**Sha48**]). *Fix $q \in (0,1)$. Let $\mathrm{num}_d(q)$ denote the minimum cardinality of a subset S of $\mathcal{X}^d$ such that $\Pr(S) \geq q$. Then*

$$\lim_{d\to\infty} \frac{\log(\mathrm{num}_d(q))}{d} = H(\bar{p}).$$

The right-hand side is independent of q; thus for large d, the ratio is nearly independent of q. Theorem 4.4.2 will follow immediately from:

Theorem 4.4.3 ([**Sha48**]). *Let $R < H(\bar{p})$ and let $S(d) \subset \mathcal{X}^d$ be a sequence of subsets with $|S(d)| \leq 2^{dR}$. Then for any $\eta > 0$, there exists d_0 such that for all $d \geq d_0$, $\Pr(S(d)) < \eta$.*

In other words, *any* sequence of subsets of size less than $2^{dH(\bar{p})}$ accumulates an arbitrarily small amount of the probability.

Proof of Theorem 4.4.3. Write

$$S(d) = S(d)_{\delta-\mathrm{small}} \sqcup S(d)_{\epsilon-\mathrm{typ}},$$

where

$$S(d)_{\delta-\mathrm{small}} = S(d) \cap \mathcal{X}^d_{\delta-\mathrm{small}} \quad \text{and} \quad S(d)_{\epsilon-\mathrm{typ}} = S(d) \cap \mathcal{X}^d_{\epsilon-\mathrm{typ}}.$$

Then $|S(d)_{\epsilon-\text{typ}}| < |S(d)| < 2^{dR}$. By (4.4.3), each element of $S(d)_{\epsilon-\text{typ}}$ has probability at most $2^{-d(H(\bar{p})-\epsilon)}$; thus

$$\Pr(S(d)_{\epsilon-\text{typ}}) \leq 2^{-d(H(\bar{p})-\epsilon)} 2^{dR}.$$

Take $\epsilon < H(\bar{p}) - R$, and d_0, δ such that $2^{-d_0(H(\bar{p})-R-\epsilon)} + \delta < \eta$; i.e., $\Pr(S(d_0)_{\epsilon-\text{typ}}) + \Pr(S(d_0)_{\delta-\text{small}}) < \eta$. □

4.4.2. Distances between probability distributions.

In the noiseless channel theorem we want to compare the distributions $\bar{p}$ and $\bar{q} := \mathcal{D} \circ \mathcal{E}(\bar{p})$, and in the noisy channel theorem we will more generally want to compare $\bar{p}$ and $\bar{q} := \Lambda_C(\bar{p})$, for an arbitrary channel Λ_C.

We have already seen the ℓ_1-norm $||\bar{x}||_1 = \sum_j |x_j|$. We can use it to measure how far $\bar{p}$ is from $\bar{q}$:

$$||\bar{p} - \bar{q}||_1 = \sum_j |p_j - q_j|.$$

Another notion of distance is fidelity: given $\bar{p}, \bar{q}$, consider the vectors

$$\sqrt{\bar{p}} := (\sqrt{p_1}, \ldots, \sqrt{p_n}), \qquad \sqrt{\bar{q}} \in S^{n-1} \subset \mathbb{R}^n,$$

where S^{n-1} denotes the unit sphere.

Definition 4.4.4. The *fidelity* of $\bar{p}$ and $\bar{q}$ is

$$F(\bar{p}, \bar{q}) := \left\langle \sqrt{\bar{p}} \,\middle|\, \sqrt{\bar{q}} \right\rangle = \left(\sum_j \sqrt{p_j q_j} \right).$$

Note that $F(\bar{p}, \bar{p}) = 1$.

We introduce fidelity because in quantum information theory, the quantum analog of fidelity will have certain computational advantages.

4.4.3. Statement of the noiseless channel theorem.

Definition 4.4.5. A channel $\Lambda_C : \mathbb{R}^{|\mathcal{X}|} \to \mathbb{R}^{|\mathcal{X}|}$ is *noiseless* if it is a permutation matrix.

In the following theorem, the noiseless channel is $\Lambda_C = \text{Id}_{\mathcal{H}}$.

Theorem 4.4.6 (Shannon's noiseless channel theorem, precise version, [**Sha48**]). *Let a source $\mathcal{X}$ have entropy $H(\bar{p})$. Let $\mathcal{Y}(d)$ be a sequence of compression spaces. If $|\mathcal{Y}(d)| > 2^{dH(\bar{p})}$, then there exist encoding and decoding channels*

$\mathcal{E}_d : \mathcal{X}^d \to \mathcal{Y}(d)$ and $\mathcal{D}_d : \mathcal{Y}(d) \to \mathcal{X}^d$ such that the encoding and decoding

$$\mathcal{X}^d \xrightarrow{\mathcal{E}_d} \mathcal{Y}(d) \xrightarrow{\mathcal{D}_d} \mathcal{X}^d$$

satisfies, as $d \to \infty$,

$$\|\mathcal{D}_d \circ \mathcal{E}_d(\bar{p}^d) - \bar{p}^d\|_1 \to 0.$$

It is not possible to reliably encode and decode if $|\mathcal{Y}(d)| < 2^{dH(\bar{p})}$.

We could instead replace the conclusion by, as $d \to \infty$,

$$F(\mathcal{D}_d \circ \mathcal{E}_d(\bar{p}^d), \bar{p}^d) \to 1.$$

The proof is clear: just transmit all the ϵ-typical sequences and discard the others.

4.5. Transmission over noisy channels

For simplicity, in this section assume $\mathcal{Y} \cong \mathcal{X}$.

4.5.1. Channels correspond to conditional probability distributions. A channel $\Lambda_C : \mathbb{R}^{|\mathcal{X}|} \to \mathbb{R}^{|\mathcal{Y}|}$ is equivalent to a conditional probability distribution $P^{\Lambda}_{\mathcal{Y}|\mathcal{X}}(y|x) := (\Lambda_C)_{y,x}$, where $(\Lambda_C)_{y,x}$ is the (y, x) entry of the matrix Λ_C.

Thus when x is sent through the channel, $P^{\Lambda}_{\mathcal{Y}|\mathcal{X}}(y|x)$ is the probability that y is received.

Given a distribution of symbols $p_{\mathcal{X}}$ on $\mathcal{X}$, we obtain a joint probability distribution $p_{\mathcal{X}\times\mathcal{Y}}$ on $\mathcal{X} \times \mathcal{Y}$ by $p_{\mathcal{X}\times\mathcal{Y}}(x, y) = p_{\mathcal{X}}(x)P^{\Lambda}_{\mathcal{Y}|\mathcal{X}}(y|x)$ whose $\mathcal{X}$-marginal distribution is $p_{\mathcal{X}}$ and whose conditional probability satisfies $p_{\mathcal{Y}|\mathcal{X}} = P^{\Lambda}_{\mathcal{Y}|\mathcal{X}}$.

It would be reasonable to expect that by introducing some redundancy, reliable transmission will be possible if for all $x \in \mathcal{X}$, $P^{\Lambda}_{\mathcal{Y}|\mathcal{X}}(x|x)$ is close to one. To quantify "some" and "close", we need the notion of conditional entropy.

4.5.2. Conditional entropy.

Definition 4.5.1. Let $\mathcal{X}, \mathcal{Y}$ be finite sets, and let $p_{\mathcal{X}\times\mathcal{Y}}$ be a probability distribution on $\mathcal{X} \times \mathcal{Y}$ with induced marginals $p_{\mathcal{X}}, p_{\mathcal{Y}}$. The *conditional entropy* is

$$H(\bar{p}_{\mathcal{Y}}|\bar{p}_{\mathcal{X}}) := -\sum_{x\in\mathcal{X}, y\in\mathcal{Y}} p_{\mathcal{X}\times\mathcal{Y}}(x, y)\log(p_{\mathcal{Y}|\mathcal{X}}(y|x)). \tag{4.5.1}$$

Conditional entropy measures the average ambiguity of the received signal.

If $p_{\mathcal{X}\times\mathcal{Y}}$ arises as in the channel set-up and there is no noise, then $H(\bar{p}_{\mathcal{Y}}|\overline{p}_{\mathcal{X}}) = 0$.

Rewrite (4.5.1) to get another interpretation of conditional entropy as follows:

$$\begin{aligned} H(\bar{p}_{\mathcal{Y}}|\bar{p}_{\mathcal{X}}) &= -\sum_{x,y} p_{\mathcal{X}\times\mathcal{Y}}(x,y)\log\frac{p_{\mathcal{X}\times\mathcal{Y}}(x,y)}{p_{\mathcal{X}}(x)} \\ &= -\sum_{x,y} p_{\mathcal{X}\times\mathcal{Y}}(x,y)\log p_{\mathcal{X}\times\mathcal{Y}}(x,y) + \sum_{x}\left(\sum_{y} p_{\mathcal{X}\times\mathcal{Y}}(x,y)\right)\log p_{\mathcal{X}}(x) \\ &= -\sum_{x,y} p_{\mathcal{X}\times\mathcal{Y}}(x,y)\log p_{\mathcal{X}\times\mathcal{Y}}(x,y) + \sum_{x} p_{\mathcal{X}}(x)\log p_{\mathcal{X}}(x). \end{aligned}$$

Thus

$$H(\bar{p}_{\mathcal{Y}}|\bar{p}_{\mathcal{X}}) = H(\bar{p}_{\mathcal{X}\times\mathcal{Y}}) - H(\bar{p}_{\mathcal{X}}). \tag{4.5.2}$$

The uncertainty of $p_{\mathcal{X}\times\mathcal{Y}}$ is the uncertainty of $p_{\mathcal{X}}$ plus the uncertainty of $p_{\mathcal{Y}}$ given $p_{\mathcal{X}}$.

In particular,

$$H(\bar{p}_{\mathcal{X}}) + H(\bar{p}_{\mathcal{Y}}) \geq H(\bar{p}_{\mathcal{X}\times\mathcal{Y}}) = H(\bar{p}_{\mathcal{X}}) + H(\bar{p}_{\mathcal{Y}}|\bar{p}_{\mathcal{X}}); \tag{4.5.3}$$

i.e.,

$$H(\bar{p}_{\mathcal{Y}}) \geq H(\bar{p}_{\mathcal{Y}}|\bar{p}_{\mathcal{X}}). \tag{4.5.4}$$

With extra knowledge, our uncertainty about $p_{\mathcal{Y}}$ cannot increase, and it decreases unless $p_{\mathcal{X}}$ and $p_{\mathcal{Y}}$ are independent.

In §4.5.6 we will need to allow entropy conditioned on two distributions; e.g.,

$$H(\bar{p}_{\mathcal{Y}}|\bar{p}_{\mathcal{X}}, \bar{p}_{\mathcal{Z}}) := H(\bar{p}_{\mathcal{X}\times\mathcal{Y}\times\mathcal{Z}}) - H(\bar{p}_{\mathcal{X}}) - H(\bar{p}_{\mathcal{Z}}). \tag{4.5.5}$$

4.5.3. Statement of results. We saw that the capacity of a noiseless channel is $H(\bar{p}_{\mathcal{X}})$. Given a channel Λ_C with noise, we must subtract off something to account for the uncertainty that a received symbol y was the symbol sent. This something is the conditional entropy.

Definition 4.5.2. Given a channel Λ_C with noise given by $P^{\Lambda}_{\mathcal{Y}|\mathcal{X}}$ and symbols sent according to $\bar{p}_{\mathcal{X}}$, define the *capacity*

$$\mathrm{Cap}(\Lambda_C, \bar{p}_{\mathcal{X}}) := H(\bar{p}_{\mathcal{X}}) - H(\bar{p}_{\mathcal{X}}|\bar{p}_{\mathcal{Y}}). \tag{4.5.6}$$

More importantly, if $P^{\Lambda}_{\mathcal{Y}|\mathcal{X}}$ is known, one can code the emitted messages before transmission to be adapted to minimize the effect of the noise:

Definition 4.5.3. The *capacity* of a channel $\Lambda_C : \mathbb{R}^{|\mathcal{X}|} \to \mathbb{R}^{|\mathcal{Y}|}$ is

$$\mathrm{Cap}(\Lambda_C) := \max_{\bar{q}_{\mathcal{X}}} \left(H(\bar{q}_{\mathcal{X}}) - H(\bar{q}_{\mathcal{X}}|\bar{p}_{\mathcal{Y}})\right), \tag{4.5.7}$$

where the maximum is taken over all probability distributions on the source.

The names are justified by the following theorems:

Theorem 4.5.4 ([**Sha48**]). *Let a discrete channel* $\Lambda_C : \mathbb{R}^{|\mathcal{X}|} \to \mathbb{R}^{|\mathcal{Y}|}$ *with probability distribution on symbols* $p_{\mathcal{X}}$ *have capacity* $\mathrm{Cap}(\Lambda_C, p_{\mathcal{X}})$. *Then reliable transmission of messages of size* 2^{dR} *is possible if* $R < \mathrm{Cap}(\Lambda_C, p_{\mathcal{X}})$, *and it is not possible if* $R > \mathrm{Cap}(\Lambda_C, p_{\mathcal{X}})$.

Theorem 4.5.5 (Shannon's noisy channel theorem, [**Sha48**]). *Let a discrete channel* $\Lambda_C : \mathbb{R}^{|\mathcal{X}|} \to \mathbb{R}^{|\mathcal{Y}|}$ *have capacity* $\mathrm{Cap}(\Lambda_C)$. *Then reliable transmission of messages of size* 2^{dR} *is possible if* $R < \mathrm{Cap}(\Lambda_C)$, *and it is not possible if* $R > \mathrm{Cap}(\Lambda_C)$.

In particular, unless $\mathrm{Cap}(\Lambda_C)$ is zero, reliable (albeit potentially slow) communication is always possible.

Proofs are given in §4.5.7.

4.5.4. Mutual information.

Using the equalities $H(p_{\mathcal{X}}|p_{\mathcal{Y}}) = H(p_{\mathcal{X}\times\mathcal{Y}}) - H(p_{\mathcal{Y}})$ and $H(p_{\mathcal{Y}}|p_{\mathcal{X}}) = H(p_{\mathcal{X}\times\mathcal{Y}}) - H(p_{\mathcal{X}})$, we have three interpretations of the rate of reliable transmission (4.5.6):

$H(\bar{p}_{\mathcal{X}}) - H(\bar{p}_{\mathcal{X}}|\bar{p}_{\mathcal{Y}})$,

the information sent minus the uncertainty of what was sent,

$= H(\bar{p}_{\mathcal{Y}}) - H(\bar{p}_{\mathcal{Y}}|\bar{p}_{\mathcal{X}})$,

the information received minus the noise,

(4.5.8)

$= H(\bar{p}_{\mathcal{X}}) + H(\bar{p}_{\mathcal{Y}}) - H(\bar{p}_{\mathcal{X}\times\mathcal{Y}})$,

the sum of the information sent and received minus the joint entropy, essentially the average bits/symbol common to what was sent and received.

The following definition is motivated by the third interpretation:

Definition 4.5.6. Given a probability distribution $p_{\mathcal{X}\times\mathcal{Y}}$ with marginals $p_{\mathcal{X}}, p_{\mathcal{Y}}$, the *mutual information*, denoted $I(\bar{p}_{\mathcal{X}} : \bar{p}_{\mathcal{Y}})$, is the quantity (4.5.8).

By Proposition 4.3.4(2), mutual information is nonnegative.

Exercise 4.5.7. Prove the *data processing inequality*: let $\mathcal{Y} = [m]$, let $S \in Mat_{m\times m}$ be doubly stochastic, and let $p_{\mathcal{X}\times\mathcal{Y}}$ be a probability distribution on $\mathcal{X}\times\mathcal{Y}$. Show that

$$I(\bar{p}_{\mathcal{X}} : \bar{p}_{\mathcal{Y}}) \geq I(\bar{p}_{\mathcal{X}} : S\bar{p}_{\mathcal{Y}}).$$

4.5.5. Toy examples.

Example 4.5.8. Say $p_{\mathcal{X}} = (q, 1-q)$ is fixed and that 0's and 1's are transmitted with each having a probability of error .01; i.e.,

$$\begin{aligned}
p_{\mathcal{X}\times\mathcal{Y}}(0,0) &= .99q, \\
p_{\mathcal{X}\times\mathcal{Y}}(0,1) &= .01q, \\
p_{\mathcal{X}\times\mathcal{Y}}(1,0) &= .99(1-q), \\
p_{\mathcal{X}\times\mathcal{Y}}(1,1) &= .01(1-q).
\end{aligned}$$

We compute

$$\begin{aligned}
&H(\bar{p}_{\mathcal{Y}}|\bar{p}_{\mathcal{X}}) \\
&= -[.99q\log(.99) + .01q\log(.01) + .99(1-q)\log(.99) + .01(1-q)\log(.01)] \\
&= -[.99\log(.99) + .01\log(.01)] \\
&\sim .081.
\end{aligned}$$

The units here are bits/symbol. Say that we can transmit 1,000 symbols/second. Then (4.5.6) predicts a transmission rate of $1{,}000(H(q,1-q)-.081)$ bits/second. It is optimal to take $q = \frac{1}{2}$, and one obtains a rate of 919 bits/second. If the probability of error goes up to $\frac{1}{2}$, then $H(\bar{p}_{\mathcal{Y}}|\bar{p}_{\mathcal{X}}) = 1$ and our discussion predicts a rate of $1{,}000(1-1) = 0$ bits/second, which agrees with our intuition that what is received is just noise.

In general, if noise acts uniformly, it will be best to take $p_{\mathcal{X}}$ as the uniform distribution.

Example 4.5.9. Let $\mathcal{X} = \{a, b, c\}$. Assume the symbol a is never affected by noise, and assume b (resp. c) has probability p of being transmitted correctly and $1-p$ of being flipped to c (resp. b). We will choose transmission probabilities $\bar{p}_{\mathcal{X}} = (p_a, p_b, p_c)$ to optimize capacity.

Given the symmetry of the situation, we should set $p_b = p_c =: p_{b,c}$. Then

$$\begin{aligned}
H(\bar{p}_{\mathcal{X}}) &= -p_a \log p_a - 2p_{b,c}\log p_{b,c}, \\
H(\bar{p}_{\mathcal{X}}|\bar{p}_{\mathcal{Y}}) &= 2p_{b,c}H(p, 1-p).
\end{aligned}$$

We want to choose $p_a, p_{b,c}$ to maximize $H(\bar{p}_{\mathcal{X}}) - H(\bar{p}_{\mathcal{Y}}|\bar{p}_{\mathcal{X}})$, subject to the constraint that $p_a + 2p_{b,c} = 1$. We have

$$H(\bar{p}_{\mathcal{X}}) - H(\bar{p}_{\mathcal{X}}|\bar{p}_{\mathcal{Y}}) = -p_a\log p_a - 2p_{b,c}\log p_{b,c} - 2p_{b,c}H(p, 1-p).$$

Viewed as a function of $p_a, p_{b,c}$, we have a standard maximization problem from calculus.

Exercise 4.5.10. Use the method of Lagrange multipliers with the constraint $p_a + 2p_{b,c} = 1$ to show that it is optimal to take

$$p_a = \frac{2^{H(p,1-p)}}{2^{H(p,1-p)}+2}, \quad p_{b,c} = \frac{1}{2^{H(p,1-p)}+2}. \tag{4.5.9}$$

Using (4.5.9), we obtain

$$H(\bar{p}_X) - H(\bar{p}_X|\bar{p}_Y) = \log \frac{2^{H(p,1-p)}+2}{2^{H(p,1-p)}}.$$

When $p = 1$, i.e., no errors, we obtain $\log(3)$, as we have a noiseless channel with three symbols. When $p = \frac{1}{2}$, we obtain $\log(2)$, as the second and third symbols are indistinguishable. We essentially have two symbols. In general, the capacity is between these two, and the first symbol is used somewhere between the same number of times and twice as often as the other two symbols.

4.5.6. Communication with the help of a correction channel. Here is an example that will give intuition into the conditional entropy $H(\bar{p}_X|\bar{p}_Y)$, which Shannon calls the "equivocation", as the amount of extra information that must be supplied to correct errors on a noisy channel.

Consider Figure 4.5.1. In the transmission from the source to the receiver, an observer is allowed to see what is transmitted and what the receiver gets. The observer is then allowed to send correction data through a noiseless channel to allow the receiver to correct the message. The question is: how much information must the observer send to the receiver to enable correction; i.e., how much capacity does the (noiseless) correction channel need to correct errors?

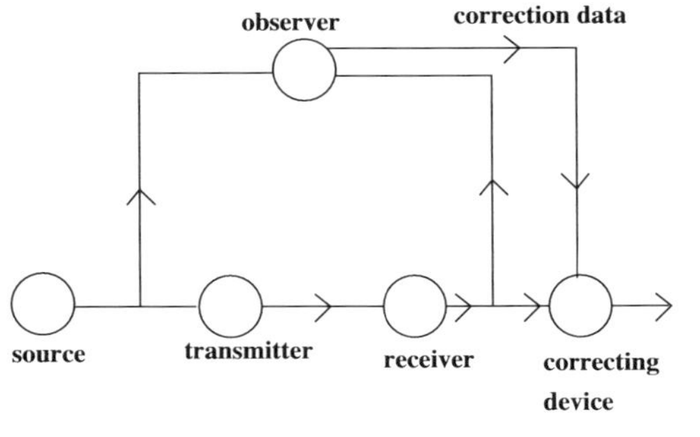

Figure 4.5.1. Observer sees corruption and corrects

Theorem 4.5.11 ([**Sha48**]). *In the scheme above errors can be reliably corrected if and only if* Cap(*correction channel*) $\geq H(\bar{p}_{\mathcal{X}}|\bar{p}_{\mathcal{Y}})$.

Proof. Assume Cap(*correction channel*) $\geq H(\bar{p}_{\mathcal{X}}|\bar{p}_{\mathcal{Y}})$. Say $\bar{y}_0^d$ is received when $\bar{x}_0^d$ is sent. By the same argument as in Theorem 4.4.1 with the set

$$\mathcal{X}^d_{\epsilon-\text{typ},\bar{y}_0} := \{\bar{x}^d \mid \Pr_{\mathcal{X}|\mathcal{Y}}(\bar{x}^d|\bar{y}_0^d) \geq 1-\delta\}$$

playing the role of $\mathcal{X}^d_{\epsilon-\text{typ}}$, there exist on the order of $2^{dH(\bar{p}_{\mathcal{X}}|\bar{p}_{\mathcal{Y}})}$ possible $\bar{x}^d$'s that could have reasonably produced $\bar{y}_0^d$. Call $\mathcal{X}^d_{\epsilon-\text{typ},\bar{y}_0}$ the *fan* of $\bar{y}_0^d$ (see Figure 4.5.2). Thus instead of one among $n^d = 2^{d\log(n)}$ possible $\bar{x}$'s that could have been sent, we are reduced to $2^{dH(\bar{p}_{\mathcal{X}}|\bar{p}_{\mathcal{Y}})}$ possible $\bar{x}$'s. The correction channel needs to send $dH(\bar{p}_{\mathcal{X}}|\bar{p}_{\mathcal{Y}})$ bits, which is possible with an ϵ-tolerance of error since the channel has capacity at least $H(\bar{p}_{\mathcal{X}}|\bar{p}_{\mathcal{Y}})$, proving this case.

For the case Cap(*correction channel*) $< H(\overline{p}_{\mathcal{X}}|\bar{p}_{\mathcal{Y}})$, we use $H(\overline{p}_{\mathcal{X}}|\bar{p}_{\mathcal{Y}}, \overline{p}_{\mathcal{Z}})$ as in (4.5.5). Informally, given $\bar{y}$ as the received message and $\bar{z}$ as the correction, how uncertain is $\bar{x}$, the original message?

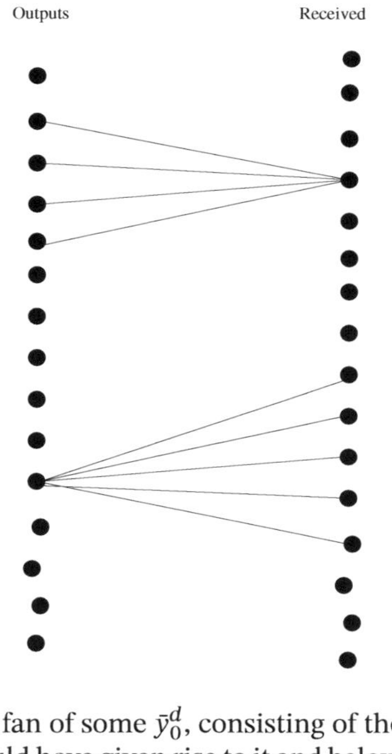

Figure 4.5.2. The fan of some $\bar{y}_0^d$, consisting of the $\sim 2^{dH(p_{\mathcal{X}}|p_{\mathcal{Y}})}$ reasonable $\bar{x}$'s that could have given rise to it and below it the fan of some $\bar{x}_1^d$ consisting of the $\sim 2^{dH(p_{\mathcal{Y}}|p_{\mathcal{X}})}$ reasonable $\bar{y}$'s that it could be corrupted to

Exercise 4.5.12. For probability distributions $p_{\mathcal{X}}, p_{\mathcal{Y}}, p_{\mathcal{Z}}$, show that

$$\begin{aligned} H(\bar{p}_{\mathcal{X}}|\bar{p}_{\mathcal{Y}}, \bar{p}_{\mathcal{Z}}) &\geq H(\bar{p}_{\mathcal{X}}|\bar{p}_{\mathcal{Y}}) - H(\bar{p}_{\mathcal{Z}}|\bar{p}_{\mathcal{Y}}) \\ &\geq H(\bar{p}_{\mathcal{X}}|\bar{p}_{\mathcal{Y}}) - H(\bar{p}_{\mathcal{Z}}). \end{aligned}$$

Now let $\bar{x}$ be the output of the source, $\bar{y}$ the received signal, and $\bar{z}$ the signal sent over the correction channel. The uncertainty of what was sent, given $\bar{y}, \bar{z}$ is $H(\bar{p}_{\mathcal{X}}|\bar{p}_{\mathcal{Y}}, \bar{p}_{\mathcal{Z}})$. Since this is bounded away from zero we cannot recover $\bar{x}$ reliably by the noiseless theorem. □

4.5.7. Proof of the noisy channel theorems. In Theorem 4.5.4, $\bar{p}_{\mathcal{X}}$ is given to us. In Theorem 4.5.5 we choose $\bar{p}_{\mathcal{X}}$ to maximize the mutual information, i.e., achieve the capacity. The argument is the same in both cases but the notation needed is slightly different. We present the notation for the proof of Theorem 4.5.5.

Proof. Choose $\bar{p}_{\mathcal{X}}$ to achieve $\text{Cap}(\Lambda_C)$. Use $\bar{p}_{\mathcal{X}}, \bar{p}_{\mathcal{Y}}$ to split $\mathcal{X}^d$ and $\mathcal{Y}^d$ into ϵ-typical and δ-small subsets with the ϵ-typical subsets, respectively, of sizes roughly (i.e., ignoring ϵ, δ) $2^{dH(\bar{p}_{\mathcal{X}})}, 2^{dH(\bar{p}_{\mathcal{Y}})}$.

For each $\bar{y} \in \mathcal{Y}^d_{\epsilon-\text{typ}}$, as in the proof of Theorem 4.5.11, there are about $2^{dH(\bar{p}_{\mathcal{X}}|\bar{p}_{\mathcal{Y}})}$ elements of $\mathcal{X}^d_{\epsilon-\text{typ}}$ that could have reasonably been sent with $\bar{y}$ received, the fan of $\bar{y}$.

Now assume $R < \text{Cap}(\Lambda_C)$ and one sends a message M^d of 2^{dR} bits. Assume that a *random* encoding of the symbols $M^d \to \mathcal{X}^d$ is chosen, so on the order of 2^{dR} bits are distributed at random among $2^{dH(\bar{p}_{\mathcal{X}})} \sim |\mathcal{X}^d_{\epsilon-\text{typ}}|$ sites. Thus (ignoring the $\pm 2^{d\epsilon}$'s) the probability of any given element of $\mathcal{X}^d_{\epsilon-\text{typ}}$ being a message is $\frac{2^{dR}}{2^{dH(\bar{p}_{\mathcal{X}})}}$, and the probability of it not being a message is $1 - 2^{d(R-H(\bar{p}_{\mathcal{X}}))}$. Thus the probability that none of the points are in the fan, other than the $\bar{x}$ that was sent (again ignoring ϵ, δ's), is

$$[1 - 2^{d(R-H(\bar{p}_{\mathcal{X}}))}]^{2^{dH(\bar{p}_{\mathcal{X}}|\bar{p}_{\mathcal{Y}})}} \tag{4.5.10}$$

where the outer exponent is approximately the number of points in the fan.

Write $R - H(p_{\mathcal{X}}) = -H(\bar{p}_{\mathcal{X}}|\bar{p}_{\mathcal{Y}}) - \eta$ for some $\eta > 0$, so (4.5.10) becomes

$$[1 - 2^{-dH(\bar{p}_{\mathcal{X}}|\bar{p}_{\mathcal{Y}}) - d\eta}]^{2^{dH(\bar{p}_{\mathcal{X}}|\bar{p}_{\mathcal{Y}})}}.$$

By Exercise 4.5.13 below, this limits to 1 as $d \to \infty$.

The case $R > \text{Cap}(\Lambda_C)$ is left to the reader. □

Exercise 4.5.13. Show that, for $H \geq 0, \eta > 0$, $\lim_{d\to\infty}[1 - 2^{-dH-d\eta}]^{2^{dH}} = 1$. ⊙

Note that the phrase "take a random encoding" is not constructive. It gives no recipe for how to do so.

Remark 4.5.14. After presenting the proof, Shannon remarks: "An attempt to obtain a good approximation to ideal coding by following the method of the proof is generally impractical. ... Probably this is no accident but is related to the difficulty of giving an explicit construction for a good approximation to a random sequence." To my knowledge, this is the first time that the difficulty of "finding hay in a haystack" (phrase due to Howard Karloff) is mentioned in print. This problem is central to complexity: for example, Valiant's algebraic version of $\mathbf{P} \neq \mathbf{NP}$ may be phrased as the problem of finding a sequence of explicit polynomials that are difficult to compute. (It is known that a random sequence is indeed difficult to compute.) According to A. Wigderson (personal communication), the difficulty of writing down random objects was explicitly discussed by Erdös in the context of random graphs, at least as early as 1947, in relation to his seminal paper [**Erd47**]. This paper, along with [**Sha48**], gave rise to the now ubiquitous probabilistic method in complexity theory, where one shows some property occurs by showing a random object has the property.

Chapter 5

Language and background material for quantum information theory

This chapter serves as preparation for the quantum information theory presented in Chapter 6. Section 5.1 presents a convenient reformulation of the postulates of quantum mechanics in terms of density operators. The reformulation gives rise to several new concepts and ways of viewing composite systems as discussed in §5.2. We then discuss distances between density operators in §5.3. A brief discussion of singular value and polar decompositions is presented in §5.4. Finally, we define quantum channels and describe several presentations of them in §5.5.

5.1. Reformulation of the postulates of quantum mechanics

We discuss two inconveniences about our formulation of the postulates of quantum mechanics which lead to a formulation of the postulates in terms of density operators.

5.1.1. Partial measurements. In §2.3.4, a measurement of a state

$$|\psi\rangle = \sum z_I |I\rangle \in (\mathbb{C}^2)^{\otimes d}$$

was defined to be a procedure that produces $I = (i_1, \ldots, i_d) \in \{0,1\}^d$ with probability $|z_I|^2$. But in our algorithms, we worked not in $(\mathbb{C}^2)^{\otimes d}$, but $(\mathbb{C}^2)^{\otimes d+m}$

where there were m "workspace" qubits we were not interested in measuring. Our measurements really were the projections onto the *spaces* $|I\rangle \otimes (\mathbb{C}^2)^{\otimes m}$. To account for this, and for other reasons that will become clear in this chapter, we define a generalized notion of measurement that allows for projection onto spaces.

To prepare for the generalization, rewrite

$$\begin{aligned} |z_I|^2 &= |\langle \psi | I \rangle|^2 \\ &= \langle \psi | I \rangle \langle I | \psi \rangle \\ &= \langle \psi | \operatorname{Proj}_I |\psi\rangle, \end{aligned}$$

where $\operatorname{Proj}_I : (\mathbb{C}^2)^{\otimes d} \to \mathbb{C}|I\rangle$ is the orthogonal projection onto the line spanned by $|I\rangle$.

Now say that we are only interested in the first d qubits of a system of $d+m$ qubits and that we want to know the probability that a measurement gives rise to some I represented by a vector $|I\rangle \in (\mathbb{C}^2)^{\otimes d}$, but we have $|\psi\rangle \in (\mathbb{C}^2)^{\otimes d+m}$. The probability of obtaining $|I\rangle$ given $|\psi\rangle$ after measuring and discarding the workspace qubits is

$$\begin{aligned} \Pr(|I\rangle \mid |\psi\rangle) &= \sum_{J \in \{0,1\}^m} \Pr(|IJ\rangle \mid |\psi\rangle) \\ &= \sum_J \langle \psi | IJ \rangle \langle IJ | \psi \rangle \\ &= \langle \psi | (|I\rangle\langle I| \otimes \operatorname{Id}_{(\mathbb{C}^2)^{\otimes m}}) |\psi\rangle \\ &= \langle \psi | \operatorname{Proj}_{\mathcal{M}} |\psi\rangle, \end{aligned}$$

where $\operatorname{Proj}_{\mathcal{M}} : (\mathbb{C}^2)^{\otimes d+m} \to |I\rangle \otimes (\mathbb{C}^2)^{\otimes m} =: \mathcal{M}$ is the orthogonal projection operator.

Previously, we could project onto any linear subspace $\mathcal{M}$ by first using a unitary transform to make $\mathcal{M}$ a coordinate subspace, then projecting onto the coordinate subspace, then taking the inverse transform. Below we will incorporate allowing projection onto an arbitrary linear subspace into the definition of a measurement.

It will be useful to have a more general notion of a projection operator beyond orthogonal projection.

Definition 5.1.1. Let V be a vector space. An element $X \in \operatorname{End}(V)$ is a *projection operator* if $X^2 = X$.

Exercise 5.1.2. Verify that our old notion of orthogonal projection in a Hilbert space, namely writing $\mathcal{H} = \mathcal{M} \oplus \mathcal{M}^\perp$, $|v\rangle = |v_1\rangle \oplus |v_2\rangle$ with $|v_1\rangle \in \mathcal{M}$, $|v_2\rangle \in \mathcal{M}^\perp$, and $\operatorname{Proj}_{\mathcal{M}}(|v\rangle) = |v_1\rangle$, satisfies Definition 5.1.1.

The following chart from [**KSV02**] illustrates the comparison between the classical (ℓ_1) and quantum (ℓ_2) probability theories so far:

	Classical (ℓ_1) probability	Quantum (ℓ_2) probability
Event	subset $M \subset \mathcal{X}$	subspace $\mathcal{M} \subset \mathcal{H}$
Probability distribution	$p : \mathcal{X} \to \mathbb{R}_{\geq 0}$ with $\lvert p \rvert_1 = 1$	$\lvert\psi\rangle \in \mathcal{H}$ with $\lvert\psi\rvert_2 = 1$
Probability	$\Pr(M) = \sum_{j \in M} p_j$	$\Pr(\mathcal{M}) = \langle\psi\rvert \operatorname{Proj}_{\mathcal{M}} \lvert\psi\rangle$

In classical probability, one has the identity

$$\Pr(M_1 \cup M_2) = \Pr(M_1) + \Pr(M_2) - \Pr(M_1 \cap M_2). \tag{5.1.1}$$

Is there a quantum analog of (5.1.1)? Is

$$\Pr(\operatorname{span}\{\mathcal{M}_1, \mathcal{M}_2\}) = \Pr(\mathcal{M}_1) + \Pr(\mathcal{M}_2) - \Pr(\mathcal{M}_1 \cap \mathcal{M}_2)?$$

The following exercise shows the answer is no:

Exercise 5.1.3. Let $\mathcal{H} = \mathbb{C}^2$, $\mathcal{M}_1 = \mathbb{C}|0\rangle$, and $\mathcal{M}_2 = \mathbb{C}(|0\rangle + |1\rangle)$. Let $|\psi\rangle = \alpha|0\rangle + \beta|1\rangle$ with $|\alpha|^2 + |\beta|^2 = 1$.

Show that $\Pr(\operatorname{span}\{\mathcal{M}_1, \mathcal{M}_2\}) \neq \Pr(\mathcal{M}_1) + \Pr(\mathcal{M}_2) - \Pr(\mathcal{M}_1 \cap \mathcal{M}_2)$. ⊙

However, if the projection operators commute, we can recover (5.1.1):

Proposition 5.1.4. *If* $\operatorname{Proj}_{\mathcal{M}_1} \operatorname{Proj}_{\mathcal{M}_2} = \operatorname{Proj}_{\mathcal{M}_2} \operatorname{Proj}_{\mathcal{M}_1}$, *then*

$$\Pr(\operatorname{span}\{\mathcal{M}_1, \mathcal{M}_2\}) = \Pr(\mathcal{M}_1) + \Pr(\mathcal{M}_2) - \Pr(\mathcal{M}_1 \cap \mathcal{M}_2).$$

In particular, if $\mathcal{M}_1 \perp \mathcal{M}_2$, *so* $\operatorname{Proj}_{\mathcal{M}_1} \operatorname{Proj}_{\mathcal{M}_2} = \operatorname{Proj}_{\mathcal{M}_2} \operatorname{Proj}_{\mathcal{M}_1} = 0$, *then* $\Pr(\operatorname{span}\{\mathcal{M}_1, \mathcal{M}_2\}) = \Pr(\mathcal{M}_1) + \Pr(\mathcal{M}_2)$.

Exercise 5.1.5. Prove Proposition 5.1.4. ⊙

A quantitative manifestation of the failure of commutativity will be given by the uncertainty principle in §5.2.5.

5.1.2. Mixing classical and quantum probability.

In classical probability, there are many natural situations where one uses the law of total probability. A typical situation is as follows: you want a cookie, but you can't make up your mind which kind, so you decide to take one at random from the cookie jar. However, when you open the cupboard, you find there are two different cookie jars, H and T, each with a different distribution of cookies represented by vectors, say $\bar{p}_H$ and $\bar{p}_T$. You decide to flip a biased coin to decide which jar.

Say your coin is biased with probability p for heads (choice H). The resulting probability distribution is

$$p\bar{p}_H + (1-p)\bar{p}_T.$$

The probability of drawing a chocolate chip (CC) cookie is $\Pr(\mathrm{CC}) = pp_H(\mathrm{CC}) + (1-p)p_T(\mathrm{CC})$.

But what should we do when our first measurement is classical and our second quantum? Given $|\psi_H\rangle = \sum z_I|I\rangle$, $|\psi_T\rangle = \sum w_J|J\rangle$, we want to make a measurement that gives CC with probability $p|z_{\mathrm{CC}}|^2 + (1-p)|w_{\mathrm{CC}}|^2$. Unfortunately $|pz_{\mathrm{CC}} + (1-p)w_{\mathrm{CC}}|^2 \neq p|z_{\mathrm{CC}}|^2 + (1-p)|w_{\mathrm{CC}}|^2$ in general. To fix this problem, we enlarge the notion of state and further modify the notion of measurement. This will just be a reformulation of convenience.

Our problem comes from having a mixture of ℓ_1 and ℓ_2 norms. The fix will be to replace $|\psi\rangle = \sum_I z_I|I\rangle$ with an object such that the ℓ_2 norm of $|\psi\rangle$ becomes an ℓ_1 norm of the new object. Consider the endomorphism $|\psi\rangle\langle\psi| = \sum_{I,J} z_I\bar{z}_J|I\rangle\langle J| \in \mathrm{End}(\mathcal{H})$. Its diagonal entries are the probabilities.

Let Proj_J denote the projection onto the line $\mathbb{C}|J\rangle$. Then

$$\mathrm{trace}(\mathrm{Proj}_J\,|\psi\rangle\langle\psi|) = |z_J|^2$$

is the desired quantity.

Now back to our cookie jars. Set

$$\rho = p|\psi_H\rangle\langle\psi_H| + (1-p)|\psi_T\rangle\langle\psi_T|$$

and observe that

$$\mathrm{trace}(\mathrm{Proj}_J\,\rho) = p|z_J|^2 + (1-p)|w_J|^2$$

as desired.

More generally, given a finite set of states $\{|\psi_1\rangle, \dots, |\psi_s\rangle\}$, where we select the k-th one classically with probability p_k, set

$$\rho = \sum_k p_k|\psi_k\rangle\langle\psi_k| \in \mathrm{End}(\mathcal{H}). \tag{5.1.2}$$

Then the probability of measuring our state in $\mathcal{M}$ is $\mathrm{trace}(\mathrm{Proj}_{\mathcal{M}}\,\rho)$.

Since each $|\psi_k| = 1$, we have $\mathrm{trace}(\rho) = 1$.

Exercise 5.1.6. Show that for all $|\eta\rangle \in \mathcal{H}$, $\langle\eta|\rho|\eta\rangle \geq 0$.

Exercise 5.1.6 motivates the following definition:

Definition 5.1.7. An element $Z \in \mathrm{End}(\mathcal{H})$ is *positive* if $\langle v|Z|v\rangle \geq 0$ for all $|v\rangle \in \mathcal{H}$.

Remark 5.1.8. In some texts, our definition of positive is called "positive semi-definite".

Exercise 5.1.9. Show that if Z is positive, then Z is Hermitian. ⊚

Exercise 5.1.10. Let $Y \in \mathrm{Hom}(\mathcal{H}_A, \mathcal{H}_B)$. Show that $YY^\dagger$ and $Y^\dagger Y$ are positive.

This discussion motivates the following definition:

Definition 5.1.11. An operator $\rho \in \mathrm{End}(\mathcal{H})$ satisfying

(1) ρ is *positive*,

(2) $\mathrm{trace}(\rho) = 1$

is called a *density operator*. Let

$$\mathcal{D}(\mathcal{H}) \subset \mathrm{End}(\mathcal{H})$$

denote the set of density operators on $\mathcal{H}$.

Exercise 5.1.12. Show that Definition 5.1.11 has no "uninvited guests"; i.e., show that all density operators are of the form (5.1.2) for some collection of ψ_j's and some probability distribution.

Density operators will replace state vectors in our reformulation of the postulates. In Proposition 5.2.7 below, we show they are just a reformulation of state vectors, albeit in a different Hilbert space. A density operator that is diagonal in the standard basis of $\mathbb{C}^d$ corresponds to a probability distribution on $\{1, \ldots, d\}$. A rank one density operator corresponds to our old notion of state. Thus the set of density operators includes both classical probability and our old notion of state.

Definition 5.1.13. Given $Z \in \mathrm{End}(\mathcal{H})$, there is an *induced action* of Z on $\mathrm{End}(\mathcal{H})$ given by $X \mapsto ZXZ^\dagger$. Adopt the notation $Z \cdot X := ZXZ^\dagger$.

Exercise 5.1.14. Show that $\mathrm{trace}(\mathrm{Proj}_{\mathcal{M}}\rho) = \mathrm{trace}(\mathrm{Proj}_{\mathcal{M}} \cdot \rho)$. ⊚

Exercise 5.1.15. Show that the set of density operators $\mathcal{D}(\mathcal{H})$ is invariant under the induced action of $\mathbf{U}(\mathcal{H})$ on $\mathrm{End}(\mathcal{H})$.

Different scenarios can lead to the same density operator. However, two scenarios giving rise to the same density operator are physically indistinguishable.

Remark 5.1.16. The set of density operators is analogous to the set of pbits (a convex set with extremal points being the basis vectors) in that it is a convex set with its extremal points being the rank one density operators. Generalizing the discussion in §2.1.1, if $\dim \mathcal{H} = n$, the set of rank one density operators is isomorphic to the quotient of the sphere S^{2n-1} by the circle.

5.1.3. Reformulation of the postulates.

Postulate 1. Associated to any isolated physical system is a Hilbert space $\mathcal{H}$, called the *state space*. The state of the system is described by a density operator $\rho \in \mathcal{D}(\mathcal{H})$.

Postulate 2. The evolution of an isolated system is described by the action of unitary operators on ρ.

Postulate 3. (Projective or von Neumann) Measurements correspond to orthogonal decompositions $\mathcal{H} = \mathcal{M}_1 \oplus \cdots \oplus \mathcal{M}_k$. The probability that ρ is measured in state $\mathcal{M}_s$, i.e., that s is obtained from the experiment, is $\text{trace}(\text{Proj}_{\mathcal{M}_s} \cdot \rho)$. After a measurement where s is obtained, the new state is $\frac{\text{Proj}_{\mathcal{M}_s} \cdot \rho}{\text{trace}(\text{Proj}_{\mathcal{M}_s} \cdot \rho)}$.

Postulate 4 is unchanged:

Postulate 4. The state space of a composite physical system is the tensor product of the Hilbert spaces: $\mathcal{H}_{AB} = \mathcal{H}_A \otimes \mathcal{H}_B$.

Projective measurements generalize the case of our old measurements where $\mathcal{M}_j$ was projection onto the j-th orthonormal basis element and our projections onto $|I\rangle \otimes (\mathbb{C}^2)^{\otimes m}$ discussed at the beginning of this section.

We will occasionally use a variant of Postulate 3:

Postulate 3′. *POVM* or *positive operator-valued measurement* corresponds to a collection of positive operators Q_s on $\mathcal{H}$ such that $\sum_{s=1}^{\ell} Q_s = \text{Id}$. The probability that ρ is measured in the state corresponding to Q_s is $\text{trace}(Q_s \rho)$.

POVMs generalize projective measurements, where the Q_s are orthogonal projection operators. Note that the Q_s in a POVM need not be orthogonal, so there can be more than $\dim \mathcal{H}$ operators. Exercise 5.2.14 will show that, like the other apparent generalizations, Postulate 3′ is just a convenience.

5.2. Remarks on, and first consequences of, the reformulation

5.2.1. Linear algebra detour. We use the following observation from linear algebra in what follows. Let V be any finite-dimensional vector space. Then $V \otimes V^*$ admits the interpretations of

- the space of linear maps $V \to V$,
- the space of linear maps $V^* \to V^*$, and
- the space of linear forms on $V^* \otimes V$.

There is a canonical linear map $V \to V$, namely the identity. It is the unique up to scale linear map invariant under the action of $GL(V)$, and we fix the scale by requiring that the determinant be one.

Thus there must be some canonical linear form $V^*\otimes V \to \mathbb{C}$ when one views Id as a linear form on $V^*\otimes V$.

Exercise 5.2.1. What is the canonical linear form? ⊚

Recall that we have a pairing $V^* \times V \to \mathbb{C}$ given by contraction.

Exercise 5.2.2. Show that the map $V^*\otimes V \to \mathbb{C}$ given by $X \mapsto \text{trace}(X)$ is the extension of the contraction map from rank one elements to all of $V^*\otimes V$.

5.2.2. Remarks on composite systems. Just as in classical probability where one is often interested in marginal distributions, in the quantum setting we have already seen several instances where one takes "marginals". We next formalize this.

The contraction map $\mathcal{H}^*\otimes\mathcal{H} \to \mathbb{C}$ on rank one elements is written $\langle v|\otimes|w\rangle \mapsto \langle v|w\rangle$. Recall the notation $|w\rangle\langle\psi| := |w\rangle\otimes\langle\psi|$.

Definition 5.2.3. For $Y \in \text{End}(\mathcal{H}_1\otimes\mathcal{H}_2) = (\mathcal{H}_1\otimes\mathcal{H}_2)\otimes(\mathcal{H}_1^*\otimes\mathcal{H}_2^*)$, define the *partial trace* $\text{trace}_{\mathcal{H}_1}(Y) \in \text{End}(\mathcal{H}_2)$ to be the image of Y under the contraction $\mathcal{H}_1\otimes\mathcal{H}_2\otimes\mathcal{H}_1^*\otimes\mathcal{H}_2^* \to \mathcal{H}_2\otimes\mathcal{H}_2^*$ given on rank one elements by

$$|v\rangle\otimes|w\rangle\otimes\langle\phi|\otimes\langle\psi| \mapsto \langle\phi|v\rangle|w\rangle\langle\psi|.$$

When Y is a density operator, we refer to the partial trace as a *quantum marginal.*

Compare the classical and quantum situations:

Classical ℓ_1 probability	Quantum ℓ_2 probability
for $\bar{p}_{\mathcal{X}_1\times\mathcal{X}_2}$, the $\mathcal{X}_1$ marginal is $p_{\mathcal{X}_1}(x) = \sum_{y\in\mathcal{X}_2} p_{\mathcal{X}_1\times\mathcal{X}_2}(x,y)$	for $\rho \in \mathcal{D}(\mathcal{H}_1\otimes\mathcal{H}_2)$, the $\mathcal{H}_1$ quantum marginal is $\rho_1 = \text{trace}_{\mathcal{H}_2}(\rho)$
for $\bar{p}_{\mathcal{X}_1\times\mathcal{X}_2}$ and $M = M_1 \times \mathcal{X}_2$, $\Pr(M) = \sum_{j\in\mathcal{X}_2} \Pr(M_1, j)$	for $\rho \in \mathcal{D}(\mathcal{H}_1\otimes\mathcal{H}_2)$ and $\mathcal{M} = \mathcal{M}_1\otimes\mathcal{H}_2$, $\Pr(\mathcal{M}) = \text{trace}_{\mathcal{H}_1\otimes\mathcal{H}_2}(\text{Proj}_{\mathcal{M}}\cdot\rho) = \text{trace}_{\mathcal{H}_1}(\text{Proj}_{\mathcal{M}_1}\cdot\text{trace}_{\mathcal{H}_2}(\rho))$

5.2.3. Pure and mixed states. We now show that the extension of states from vectors to density operators is just a convenience.

Definition 5.2.4. Let $\rho \in \mathcal{D}(\mathcal{H})$. If $\text{rank}(\rho) = 1$, then ρ is called a *pure state,* and otherwise it is called a *mixed state.*

In Definition 5.2.4, $\text{rank}(\rho)$ is the usual rank as an endomorphism, i.e., $\dim(\rho(\mathcal{H}))$.

Exercise 5.2.5. Show that a rank one density operator is of the form $|\xi\rangle\langle\xi|$ (as opposed to $|\xi\rangle\langle\psi|$).

Exercise 5.2.6. Show that $\rho \in \mathcal{D}(\mathcal{H})$ is pure if and only if $\rho \circ \rho = \rho$.

The partial trace of a pure state can be a mixed state. For example, if $\rho = |\mathrm{epr}\rangle\langle\mathrm{epr}|$ with $|\mathrm{epr}\rangle = \frac{1}{\sqrt{2}}(|00\rangle + |11\rangle) \in \mathcal{H}_1 \otimes \mathcal{H}_2$, then

$$\mathrm{trace}_{\mathcal{H}_2}(\rho) = \frac{1}{2}(|0\rangle\langle 0| + |1\rangle\langle 1|),$$

which is mixed.

The following proposition shows that we could avoid density operators altogether by working on a larger space:

Proposition 5.2.7. *An arbitrary density operator $\rho \in \mathcal{D}(\mathcal{H})$ may be represented as the partial trace of a pure state in* $\mathrm{End}(\mathcal{H} \otimes \mathcal{H}')$ *for some Hilbert space $\mathcal{H}'$. In fact, one may take $\mathcal{H}' = \mathcal{H}^*$.*

Exercise 5.2.8. Show that given a density operator $\rho \in \mathrm{End}(\mathcal{H})$, there is a well-defined positive operator $\sqrt{\rho} \in \mathrm{End}(\mathcal{H})$ such that $\rho = \sqrt{\rho} \circ \sqrt{\rho}$. ⊚

Proof of Proposition 5.2.7. We are given $\rho \in \mathcal{H} \otimes \mathcal{H}^*$. Consider

$$|\sqrt{\rho}\rangle\langle\sqrt{\rho}| \in \mathrm{End}(\mathcal{H} \otimes \mathcal{H}^*)$$

and note that it has trace one, so it is a density operator. Then

$$\rho = \mathrm{trace}_{\mathcal{H}^*}(|\sqrt{\rho}\rangle\langle\sqrt{\rho}|) \qquad \square.$$

Exercise 5.2.9. Verify $\rho = \mathrm{trace}_{\mathcal{H}^*}(|\sqrt{\rho}\rangle\langle\sqrt{\rho}|)$.

Definition 5.2.10. A pure state whose partial trace is ρ is called a *purification* of ρ. The purification $|\sqrt{\rho}\rangle\langle\sqrt{\rho}|$ is called the *standard purification.*

Remark 5.2.11. A computational motivation for the use of density operators is as follows: it is often the case that the full Hilbert space is the tensor product of many spaces: $\mathcal{H} = \mathcal{H}_1 \otimes \cdots \otimes \mathcal{H}_D$ for some large D. However, the only Hilbert spaces relevant for the experiment are the first few, say $\mathcal{H}_1 \otimes \mathcal{H}_2$; i.e., any action taken will be the identity on the remaining spaces. Then it makes sense computationally to introduce the density operator

$$\rho = (\mathrm{Id}_{\mathcal{H}_1 \otimes \mathcal{H}_2} \otimes \mathrm{trace}_{\mathcal{H}_3 \otimes \cdots \otimes \mathcal{H}_D})|\psi\rangle\langle\psi|$$

to reduce to a computation in a much smaller Hilbert space.

Exercise 5.2.12. Show that if $|\psi\rangle\langle\psi|, |\xi\rangle\langle\xi| \in \mathrm{End}(\mathcal{H} \otimes \mathcal{H}')$ are purifications of ρ, then $|\psi\rangle = (\mathrm{Id}_{\mathcal{H}} \otimes U)|\xi\rangle$ for some $U \in \mathbf{U}(\mathcal{H}')$.

Exercise 5.2.13. Show that in a purification, one may take $\dim \mathcal{H}' = \operatorname{rank} \rho$.

Exercise 5.2.14. Show that the extension of projective measurements to POVMs is merely a convenience; i.e., the effect of any POVM may be attained by performing a projective measurement on a larger Hilbert space. ⊚

5.2.4. Quantum mechanics from probability via four properties. Hardy [**Har01**] and Shack [**Sch03**] proved that any theory satisfying four natural properties of probability must be classical probability expressed as stochastic matrices: (i) learning about a space via a minimal number of measurements, (ii) behavior of systems under specialization, (iii) behavior of composite systems, and (iv) the ability to transform from one extremal state (i.e., distribution concentrated at a point) to another. They also showed that if one adds the requirement that any such transform can be written as a product of transforms that are arbitrarily close to the identity transform, one obtains the axioms of quantum mechanics (and the minimal number of measurements to learn about the space as a function of its dimension grows). Here the reversible transformations are the elements of $\mathbf{U}(\mathcal{H})$. Originally Hardy had the extra axiom that the law of large numbers was satisfied, then Shack showed that, just like in the classical case, the law of large numbers is a consequence of the other axioms.

5.2.5. Quantum random variables and the uncertainty principle. What is the quantum analog of a random variable $X : \mathcal{X} \to \mathbb{R}$?

Say we have an orthogonal direct sum decomposition $\mathcal{H} = \mathcal{M}_1 \oplus \cdots \oplus \mathcal{M}_k$; we have an experiment with k possible outcomes. Whatever a quantum random variable is, it should give rise to a map $X : \{1, \ldots, k\} \to \mathbb{R}$. Write $\lambda_j = X(j)$; then the probability distribution induced by X must take the value λ_j with probability equal to the probability of drawing $\mathcal{M}_j$, i.e., $\mathrm{trace}(\mathrm{Proj}_{\mathcal{M}_j} \cdot \rho)$.

Exercise 5.2.15. Prove the *spectral decomposition theorem* for Hermitian operators: Hermitian operators are diagonalizable and the eigenspaces of a Hermitian operator X are orthogonal. In particular, we may write $X = \sum_\lambda \lambda P_\lambda$ where the λ are the eigenvalues of X and the P_λ are commuting projection operators: $P_\lambda P_\mu = P_\mu P_\lambda$ and $P_\lambda^2 = P_\lambda$. ⊚

The spectral decomposition theorem and the discussion above it suggest that the quantum analog of a random variable should be a Hermitian operator. When viewed in this way, a Hermitian operator is called an *observable*.

A special case of an observable is the case $\lambda_j = j$. Since λ_j and j just serve to tell us which outcome of the measurement occurred, they do not change the physics or the mathematics of the situation. Thus some texts define von Neumann measurements by:

Postulate 3 (Alternative phrasing). (Projective or von Neumann) Measurements correspond to Hermitian operators $X \in \mathrm{End}(\mathcal{H})$. Associated to X is an orthogonal direct sum decomposition $\mathcal{H} = \oplus_j \mathcal{M}_j$, where the $\mathcal{M}_j$ are the eigenspaces of X. Let λ_j denote the associated eigenvalues. The probability that ρ is measured in state $\mathcal{M}_j$, i.e., that the value λ_j is obtained from the experiment, is $\text{trace}(\text{Proj}_{\mathcal{M}_j} \cdot \rho)$. After such a (partial) measurement where λ_j is obtained, the new state is $\frac{\text{Proj}_{\mathcal{M}_j} \cdot \rho}{\text{trace}(\text{Proj}_{\mathcal{M}_j} \cdot \rho)}$.

This phrasing is convenient for physical situations where the device performing the measurement returns some natural (for the experiment) set of values.

Proposition 5.2.16. *If a system is in state ρ, the expectation of a Hermitian operator $X \in \text{End}(\mathcal{H})$ is $\text{trace}(X\rho)$.*

Proof. Write X in terms of its eigenvalues λ_j and eigenspaces $\mathcal{M}_j$. Then

$$\begin{aligned} E[X] &= \sum_j X(j) \Pr(X = j) \\ &= \sum_j \lambda_j \, \text{trace}(\text{Proj}_{\mathcal{M}_j} \rho) \\ &= \text{trace}\left(\left(\sum_j \lambda_j \, \text{Proj}_{\mathcal{M}_j}\right) \rho\right) \\ &= \text{trace}(X\rho). \end{aligned}$$

□

As with classical probability, the variance of a Hermitian operator is

$$\begin{aligned} \text{var}(X) :&= E[(X - E[X])^2] \\ &= \text{trace}((X - \text{trace}(X\rho)\,\text{Id})^2 \rho), \end{aligned}$$

and the *standard deviation* is

$$\sigma(X) = \sqrt{\text{var}(X)}.$$

One way mathematicians describe the famous Heisenberg uncertainty principle is that it is impossible to localize both a function and its Fourier transform. Another interpretation comes from probability:

Say we have two Hermitian operators X, Y. If we measure with respect to X and then measure with respect to Y, the outcome may not be the same as if we first measure with respect to Y and then measure with respect to X. The uncertainty principle will quantify this difference in terms of the commutator $[X, Y] := XY - YX$.

Proposition 5.2.17 (Heisenberg uncertainty principle). *Let X, Y be Hermitian operators, corresponding to observables on a system in state ρ. Then the standard deviations $\sigma(X), \sigma(Y)$ satisfy*

$$\sigma(X)\sigma(Y) \geq \frac{|\,\text{trace}([X, Y]\rho)|}{2}.$$

The uncertainty principle says that the failure of two Hermitian operators to commute bounds from below the product of their uncertainties. In particular, if they do not commute, neither can give rise to a classical (certain) measurement.

Exercise 5.2.18. For $X, Y \in \mathrm{Herm}(\mathcal{H})$ and $\rho \in \mathcal{D}(\mathcal{H})$, show that $\mathrm{trace}([X, Y]\rho) \in i\mathbb{R}$. ⊙

Exercise 5.2.19. For $Z \in \mathrm{End}(\mathcal{H})$ and $\rho \in \mathcal{D}(\mathcal{H})$, show that $\mathrm{trace}(Z^\dagger Z\rho) \geq 0$. ⊙

Proof of Proposition 5.2.17. Given a random variable, or Hermitian operator X and a system in state ρ, define an operator of mean zero

$$\hat{X} := X - E[X]\,\mathrm{Id}.$$

Note that $\mathrm{var}(\hat{X}) = \mathrm{var}(X)$ and $[X, Y] = [\hat{X}, \hat{Y}]$, so in what follows, we may assume X, Y have expectation zero.

By Exercise 5.2.19, for all $\lambda \in \mathbb{R}$,

$$\begin{aligned} 0 &\leq \mathrm{trace}((X + i\lambda Y)^\dagger (X + i\lambda Y)\rho) \\ &= \mathrm{trace}(X^2\rho) + \lambda i\,\mathrm{trace}([X, Y]\rho) + \lambda^2\,\mathrm{trace}(Y^2\rho) \\ &= \sigma(X)^2 + \lambda i\,\mathrm{trace}([X, Y]\rho) + \lambda^2\sigma(Y)^2. \end{aligned} \tag{5.2.1}$$

Without loss of generality assume $Y \neq 0$ (otherwise the inequality is trivial); then (5.2.1) is a quadratic function in λ which is nonnegative, so its discriminant must be nonnegative; i.e.,

$$|\,\mathrm{trace}([X, Y]\rho)|^2 - 4\sigma(X)^2\sigma(Y)^2 \geq 0.$$

□

5.2.6. Summary of classical v. quantum probabilities. Here is a summary of the correspondence between classical and quantum probability:

	Classical (ℓ_1) probability	Quantum (ℓ_2) probability
Event	subset $M \subset \mathcal{X}$	subspace $\mathcal{M} \subset \mathcal{H}$
Probability distribution	$p : \mathcal{X} \to \mathbb{R}_{\geq 0}$ with $\lvert p\rvert_1 = 1$	$\rho \in \mathcal{D}(\mathcal{H})$
Probability	$\Pr(M) = \sum_{j \in M} p_j$	$\Pr(\mathcal{M}) = \mathrm{trace}(\mathrm{Proj}_{\mathcal{M}} \cdot \rho)$
Marginals	Given $p_{\mathcal{X}\times\mathcal{Y}}$, $p_{\mathcal{X}}(x) := \sum_{y \in \mathcal{Y}} p_{\mathcal{X}\times\mathcal{Y}}(x, y)$	Given $\rho_{AB} \in \mathcal{D}(\mathcal{H}_A \otimes \mathcal{H}_B)$, $\rho_A := \mathrm{trace}_{\mathcal{H}_B}(\rho_{AB}) \in \mathcal{D}(\mathcal{H}_A)$
Random variables and expectation	$X : \mathcal{X} \to \mathbb{R}$ $E[X] = \sum_{x \in \mathcal{X}} X(x) p_{\mathcal{X}}(x)$	$X \in \mathrm{Herm}(\mathcal{H})$ $E[X] = \mathrm{trace}(X\rho)$

5.3. Distances between density operators

We now discuss the quantum analogs of the ℓ_1 norm and fidelity of §4.4.2.

Recall that if $\rho \in \mathcal{D}(\mathbb{C}^n)$ is diagonal, it corresponds to a classical probability distribution, so whatever distances and "fidelity" we define on density operators, we will require them to specialize to the ℓ_1 distance and fidelity when the density operators measured are classical. The quantum cousin of the ℓ_1 norm is the *trace* norm:

Definition 5.3.1. The *trace norm* or *Hilbert Schmidt norm* of $A \in \mathrm{Hom}(\mathcal{H}_B, \mathcal{H}_C)$ is

$$||A||_1 := \mathrm{trace}(\sqrt{A^\dagger A}).$$

Exercise 5.3.2. Show that $||A||_1$ is indeed a norm.

Exercise 5.3.3. Show that given $A \in \mathrm{End}(\mathcal{H}_1)$ and $B \in \mathrm{End}(\mathcal{H}_2)$, for $A \otimes B \in \mathrm{End}(\mathcal{H}_1 \otimes \mathcal{H}_2)$, one has

$$||A \otimes B||_1 = ||A||_1 ||B||_1.$$

Now for the quantum version of fidelity:

Definition 5.3.4. For density operators ρ, σ the *fidelity* is

$$F(\rho, \sigma) := ||\sqrt{\rho}\sqrt{\sigma}||_1.$$

Note that $F(\rho, \sigma) = F(\sigma, \rho)$.

Exercise 5.3.5. Show that if $\rho = |\phi\rangle\langle\phi|$ and $\sigma = |\psi\rangle\langle\psi|$ are pure, then $F(\rho, \sigma) = |\langle\psi|\phi\rangle|$.

Exercise 5.3.6. Show that $F(\rho, \sigma) = (\mathrm{trace}(\sqrt{\sqrt{\rho}\sigma\sqrt{\rho}}))$.

Exercise 5.3.7. Show that $F(\rho_1 \otimes \rho_2, \sigma_1 \otimes \sigma_2) = F(\rho_1, \sigma_1)F(\rho_2, \sigma_2)$ where $\rho_j, \sigma_j \in \mathcal{D}(\mathcal{H}_j)$, $j = 1, 2$.

If ρ, σ are classical (i.e., diagonal in the standard basis), this fidelity agrees with the classical fidelity.

5.4. Singular value and polar decompositions

Definition 5.4.1. For any linear map $f : \mathcal{H}_1 \to \mathcal{H}_2$, its *singular values* are the nonnegative square roots of the eigenvalues of the positive operator $f^\dagger f$. Here $f^\dagger : \mathcal{H}_2 \to \mathcal{H}_1$ is the induced map using the Hilbert space/dual space identification of the map (2.1.7) on both spaces.

Proposition/Definition 5.4.2. *Given a matrix $X \in \mathbb{C}^n \otimes \mathbb{C}^m$, one has the decomposition*

$$X = VDU \tag{5.4.1}$$

with $V \in \mathbf{U}(n)$, $U \in \mathbf{U}(m)$, and $D \in \mathbb{C}^n \otimes \mathbb{C}^m$ is a matrix that has the singular values of X along the main diagonal and that is zero elsewhere. The decomposition (5.4.1) *is called a singular value decomposition.*

Proof. Consider the special case $n = m = \operatorname{rank}(X)$.

Recall from Exercise 5.1.10 that $X^\dagger X \in \operatorname{End}(\mathbb{C}^n)$ is positive. Thus $X^\dagger X = UDU^{-1}$ for some unitary transformation $U \in \mathbf{U}(n)$ and diagonal matrix D with positive entries. The transformation $D^{-\frac{1}{2}}$ is well-defined, where we take the positive square roots. Let $V = XUD^{-\frac{1}{2}}$, so $X = VD^{\frac{1}{2}}U^{-1}$. The matrix V is unitary as

$$\begin{aligned} V^\dagger V &= (D^{-\frac{1}{2}}U^{-1}X^\dagger)(XUD^{-\frac{1}{2}}) \\ &= \operatorname{Id}, \end{aligned}$$

proving the special case.

Exercise 5.4.3. Prove the general case by splitting

$$\mathbb{C}^n = \operatorname{Image}(X_{BA}) \oplus \operatorname{Image}(X_{BA})^\perp$$

and

$$\mathbb{C}^m = \operatorname{Image}(X_{AB}) \oplus \operatorname{Image}(X_{AB})^\perp$$

where $X_{BA} : (\mathbb{C}^m)^* \to \mathbb{C}^n$ and $X_{AB} : (\mathbb{C}^n)^* \to \mathbb{C}^m$ are the linear maps induced by X. □

One also has the related *polar decomposition:*

Proposition/Definition 5.4.4. *Given a matrix $A \in \operatorname{End}(\mathbb{C}^n)$, there exists $U \in \mathbf{U}(\mathcal{H})$ such that*

$$A = U\sqrt{A^\dagger A} = \sqrt{AA^\dagger}U. \tag{5.4.2}$$

If A is invertible, U is unique. A decomposition of the form (5.4.2) *is called a polar decomposition.*

Proof. Write the singular value decomposition of A as $A = VDW$. Set $P = W^{-1}DW$ and $U = VW$. Then $A = UP$ and

$$\begin{aligned} A^\dagger A &= (W^{-1}D^\dagger V^{-1})(VDW) \\ &= P^2, \end{aligned}$$

where to obtain the last line, recall $D^\dagger = D$.

Exercise 5.4.5. Derive the second equality in (5.4.2) and the uniqueness. □

Proposition 5.4.6. *Let $|\psi\rangle \in \mathcal{H}_A \otimes \mathcal{H}_B$. There exist orthonormal bases $|i_A\rangle$ of $\mathcal{H}_A$ and $|j_B\rangle$ of $\mathcal{H}_B$ such that $|\psi\rangle = \sum_k \lambda_k |k_A\rangle \otimes |k_B\rangle$ for some constants $\lambda_k \in \mathbb{C}$.*

Proof. Fix background orthonormal bases $|s\rangle$ of $\mathcal{H}_A$, $|y\rangle$ of $\mathcal{H}_B$ so $|\psi\rangle = \sum_{s,y} c_{s,y} |s\rangle \otimes |y\rangle$. Take the singular value decomposition of $C = (c_{s,y})$, so $C = UDV$ with $U \in \mathbf{U}(\mathcal{H}_A)$, $V \in \mathbf{U}(\mathcal{H}_B)$, and D diagonal. Set $|i_A\rangle = \sum_s U_{is}|s\rangle$ and $|j_B\rangle = \sum_y V_{jy}|y\rangle$ to conclude. □

5.5. Quantum channels

5.5.1. What is a quantum channel? We defined a classical channel to be a stochastic matrix. In the quantum world we need a definition that encompasses the operations corresponding to physical changes in the system such as the steps of quantum computation, the effects of noise, and the processing of quantum information.

What should a "quantum channel" that takes density operators on a Hilbert space $\mathcal{H}_A$ to density operators on a Hilbert space $\mathcal{H}_B$ be? To be consistent with the postulates of quantum mechanics it should come from a linear operator

$$\Lambda : \mathrm{End}(\mathcal{H}_A) \to \mathrm{End}(\mathcal{H}_B).$$

Let's make a wish list for the kinds of maps we would like to allow.

Based on our experience with quantum algorithms ("workspace qubits"), we would like to allow coupling with an auxiliary system; i.e.,

$$\rho \mapsto \rho \otimes \sigma \in \mathrm{End}(\mathcal{H}_A \otimes \mathcal{H}_C). \tag{5.5.1}$$

We would then like to allow the state $\rho \otimes \sigma$ to evolve in $\mathrm{End}(\mathcal{H}_A \otimes \mathcal{H}_C)$, i.e., be acted upon by an arbitrary $U \in \mathbf{U}(\mathcal{H}_A \otimes \mathcal{H}_C)$.

We next would like for Alice to be able to transmit qubits to Bob, i.e., to allow isometries from $\mathcal{H}_A \otimes \mathcal{H}_C$ to $\mathcal{H}_B \otimes \mathcal{H}_C$.

Finally we would like to allow measurements, in particular tracing out the $\mathcal{H}_C$ part.

In summary, we would like to allow maps of the form

$$\rho \mapsto \mathrm{trace}_{\mathcal{H}_C} \Phi_{AC,BC}(U(\rho \otimes \sigma)U^{-1}), \tag{5.5.2}$$

where $\Phi_{AC,BC} : \mathcal{H}_A \otimes \mathcal{H}_C \to \mathcal{H}_B \otimes \mathcal{H}_C$ is an isometry.

Remark 5.5.1. We will also want to allow different orders, e.g., have the unitary act after the isometry (as in "quantum teleportation"). The reader may check that this will not change our wish list.

These maps may be viewed as the discrete time evolution of a quantum system that is acted on by outside systems.

Now that we have our wish list, we consider what restrictions must be imposed and then check if we may obtain everything on the wish list.

To take density operators to density operators, a quantum channel must be *positive* and *trace preserving*; i.e., if $\rho \geq 0$, then $\Lambda(\rho) \geq 0$ and $\text{trace}(\Lambda(\rho)) = \text{trace}(\rho)$. Since we want to be able to couple our space with an auxiliary space, we must consider maps of the form

$$\Lambda \otimes \text{Id}_C : \text{End}(\mathcal{H}_A \otimes \mathcal{H}_C) \to \text{End}(\mathcal{H}_B \otimes \mathcal{H}_C),$$

and this also must be positive and trace preserving. Trace preserving is automatic, but positivity is not:

Exercise 5.5.2. Show that for $n \geq 2$ the map

$$\tau : \text{End}(\mathbb{C}^n) \to \text{End}(\mathbb{C}^n)$$
$$\sum X^i_j |i\rangle\langle j| \mapsto \sum X^j_i |i\rangle\langle j|$$

is positive, but $\tau \otimes \text{Id}_{\mathbb{C}^n} : \text{End}(\mathbb{C}^n \otimes \mathbb{C}^n) \to \text{End}(\mathbb{C}^n \otimes \mathbb{C}^n)$ is not positive. ⊚

To fix this problem, we make a definition:

Definition 5.5.3. A linear map $\Lambda : \mathcal{H}_A \to \mathcal{H}_B$ is *completely positive* if $\Lambda \otimes \text{Id}_{\text{End}(\mathcal{H}_C)}$ is positive for all $\mathcal{H}_C$.

Exercise 5.5.4. Show that to test if Λ is completely positive, it is enough to couple with $\text{Id}_{\mathbb{C}^n}$ where $n = \dim \mathcal{H}_A$. ⊚

Thus, at most we should allow *completely positive trace-preserving maps* (CPTP).

Exercise 5.5.5. Show that any isometry $f : \mathbb{C}^n \to \mathbb{C}^m$ may be viewed as the restriction of an element of $\mathbf{U}(n+m)$ to $\mathbb{C}^n$ and then projecting $\mathbb{C}^{n+m} \to \mathbb{C}^m$. I.e., there exists $U \in \mathbf{U}(n+m)$ such that $f = \text{Proj}_{\mathbb{C}^m}(U|_{\mathbb{C}^n \subset \mathbb{C}^{n+m}})$.

Exercise 5.5.6. Show that any map of the form (5.5.2) is indeed CPTP. ⊚

Thanks to Exercise 5.5.6, we are in the fortunate situation that the set of maps we would like to permit is contained in the set we must restrict to. We will soon see that they coincide.

Definition 5.5.7. A *quantum channel* between Hilbert spaces $\mathcal{H}_A$ and $\mathcal{H}_B$ is a completely positive trace-preserving (CPTP) linear map

$$\Lambda : \text{End}(\mathcal{H}_A) \to \text{End}(\mathcal{H}_B).$$

Remark 5.5.8. Exercise 5.5.5 partially addresses the classical v. quantum worlds. With more care, the outside measurement may also be viewed from a quantum perspective, where the "quantum-ness" of the measuring device is manifested on a much larger time scale so that it does not affect the experiment; see [**Pre**, Chap. 3].

Based on our definition of a quantum channel, it is not transparent if a given linear map $\Lambda : \mathrm{End}(\mathcal{H}_A) \to \mathrm{End}(\mathcal{H}_B)$ is completely positive or trace preserving. In the next two subsections we present three descriptions of quantum channels. With the first, it is easy to check complete positivity and trace preserving. With the second and third, complete positivity is automatic and it is easy to check trace preserving.

5.5.2. Choi representation. Choi [**Cho75**] had the idea that instead of studying maps $\Lambda : \mathrm{End}(\mathcal{H}_A) \to \mathrm{End}(\mathcal{H}_B)$, one could take advantage of the following tautological isomorphism to obtain an element of $\mathrm{End}(\mathcal{H}_A^* \otimes \mathcal{H}_B)$:

For finite-dimensional vector spaces V, W, there are tautological isomorphisms:

$$\begin{aligned}\mathrm{Hom}(\mathrm{End}(V), \mathrm{End}(W)) &= (W \otimes W^*) \otimes (V \otimes V^*)^* \\ &= W \otimes W^* \otimes V^* \otimes V \\ &= (W \otimes V^*) \otimes (W \otimes V^*)^* \\ &= \mathrm{End}(\mathrm{Hom}(V, W)) \\ &= \mathrm{End}(W \otimes V^*).\end{aligned}$$

We will call the isomorphism

$$\mathrm{Hom}(\mathrm{End}(\mathcal{H}_A), \mathrm{End}(\mathcal{H}_B)) \to \mathrm{End}(\mathrm{Hom}(\mathcal{H}_A, \mathcal{H}_B))$$

the *canonical Choi-Jamiołkowski isomorphism*, and we denote it

$$\mathrm{canonCJ} : \mathrm{Hom}(\mathrm{End}(\mathcal{H}_A), \mathrm{End}(\mathcal{H}_B)) \to \mathrm{End}(\mathcal{H}_B \otimes \mathcal{H}_A^*).$$

For the rest of this section, adopt the index ranges

$$\begin{aligned}1 \le i, j, k, \ell \le \dim \mathcal{H}_A, \\ 1 \le s, t, u \le \dim \mathcal{H}_B.\end{aligned}$$

Write $\{|i\rangle\}$ for a basis of $\mathcal{H}_A$ and $\{|s\rangle\}$ for a basis of $\mathcal{H}_B$.

Recall the notations $|xy\rangle := |x\rangle \otimes |y\rangle$, $\langle xy| := \langle x| \otimes \langle y|$, and $|x\rangle\langle y| := |x\rangle \otimes \langle y|$.

In terms of indices, if

$$\begin{aligned}\Lambda = \sum_{i,j,s,t} \Lambda^{\binom{i}{j}}_{\binom{s}{t}} (|t\rangle\langle s|) \otimes (\langle i| \otimes |j\rangle) \\ \in \mathrm{Hom}(\mathrm{End}(\mathcal{H}_A), \mathrm{End}(\mathcal{H}_B)) = (\mathcal{H}_B \otimes \mathcal{H}_B^*) \otimes (\mathcal{H}_A^* \otimes \mathcal{H}_A),\end{aligned}$$

then canonCJ is just the permutation of factors $(2, 3)$,

$$\mathrm{canonCJ}(\Lambda) = \sum_{i,j,s,t} \Lambda^{\binom{i}{j}}_{\binom{s}{t}} (|t\rangle\langle i|) \otimes (\langle s| \otimes |j\rangle) \in (\mathcal{H}_B \otimes \mathcal{H}_A^*) \otimes (\mathcal{H}_B^* \otimes \mathcal{H}_A).$$

This turns out to be useful for checking that Λ is trace preserving; namely Λ is trace preserving if and only if $\operatorname{trace}_{\mathcal{H}_B} \operatorname{canonCJ}(\Lambda) = \mathrm{Id}_{\mathcal{H}_A^*}$. However, total positivity only manifests itself transparently after a modification. Because of this, one incorporates the modification into the map and makes the following definition:

Definition 5.5.9. Fix an isometry $\tau : \mathcal{H}_A^* \to \mathcal{H}_A$, e.g., by fixing an orthonormal basis $|k\rangle$ of $\mathcal{H}_A$ and then mapping $\langle k| \mapsto |k\rangle$. Let $\sigma_{(1243)}$ denote the cyclic permutation of factors. The (standard) *Choi-Jamiołkowski isomorphism* is

$$\begin{array}{rcl} \mathrm{CJ} : \mathrm{Hom}(\mathrm{End}(\mathcal{H}_A), \mathrm{End}(\mathcal{H}_B)) & \longrightarrow & \mathrm{End}(\mathcal{H}_A \otimes \mathcal{H}_B) \\ \| & & \| \\ (\mathcal{H}_B \otimes \mathcal{H}_B^*) \otimes (\mathcal{H}_A^* \otimes \mathcal{H}_A) & \to & (\mathcal{H}_A \otimes \mathcal{H}_B) \otimes (\mathcal{H}_A^* \otimes \mathcal{H}_B^*) \end{array}$$

$$\Lambda \longmapsto \sigma_{(1243)} \circ (\mathrm{Id}_{\mathcal{H}_B} \otimes \mathrm{Id}_{\mathcal{H}_B^*} \otimes \tau \otimes \tau^{-1})(\Lambda).$$

In indices, if $\Lambda = \sum_{i,j,s,t} \Lambda^{\binom{i}{j}}_{\binom{s}{t}} |t\rangle\langle s| \otimes (\langle i| \otimes |j\rangle)$, then

$$\mathrm{CJ}(\Lambda) = \sum_{i,j,s,t} \Lambda^{\binom{i}{j}}_{\binom{s}{t}} |it\rangle \otimes \langle js|.$$

This is both a linear isomorphism and an isometry.

Its inverse is

$$\mathrm{CJ}^{-1} : \mathrm{End}(\mathcal{H}_A \otimes \mathcal{H}_B) \to \mathrm{Hom}(\mathrm{End}(\mathcal{H}_A), \mathrm{End}(\mathcal{H}_B))$$

$$\Sigma \mapsto (\mathrm{Id}_{\mathcal{H}_B} \otimes \mathrm{Id}_{\mathcal{H}_B^*} \otimes \tau^{-1} \otimes \tau) \circ \sigma_{(1342)}(\Sigma).$$

Just as with the canonical Choi-Jamiołkowski isomorphism, trace preserving is easy to characterize:

Proposition 5.5.10. *Λ is trace preserving if and only if* $\operatorname{trace}_{\mathcal{H}_B}(\mathrm{CJ}(\Lambda)) = \mathrm{Id}_{\mathcal{H}_A}$.

Proof. Write

$$\Lambda = \sum_{i,j,s,t} \Lambda^{\binom{i}{j}}_{\binom{s}{t}} |t\rangle\langle s| \otimes (\langle i| \otimes |j\rangle).$$

The maps $|k\rangle\langle\ell|$ form a basis of $\mathrm{End}(\mathcal{H}_A)$. Since $\operatorname{trace}(|k\rangle\langle\ell|) = \delta_{k\ell}$, we need $\operatorname{trace}(\Lambda(|k\rangle\langle\ell|)) = \delta_{k\ell}$. Now we check: $\Lambda(|k\rangle\langle\ell|) = \sum_{s,t} \Lambda^{\binom{k}{\ell}}_{\binom{s}{t}} |t\rangle\langle s|$ and $\operatorname{trace}(\Lambda(|k\rangle\langle\ell|)) = \sum_s \Lambda^{\binom{k}{\ell}}_{\binom{s}{s}}$, which is $\delta_{k\ell}$ for all k, ℓ if and only if Λ is trace preserving. On the other hand,

$$\operatorname{trace}_{\mathcal{H}_B}(\mathrm{CJ}(\Lambda)) = \sum_{ij} \left(\sum_s \Lambda^{\binom{i}{j}}_{\binom{s}{s}} \right) |j\rangle\langle i|.$$

The term in parenthesis is $\operatorname{trace}(\Lambda(|i\rangle\langle j|))$, which is δ_{ij} if and only if Λ is trace preserving, in which case we indeed obtain $\mathrm{Id}_{\mathcal{H}_A} = \sum_j |j\rangle\langle j|$. □

Eventually we will see Λ is completely positive if and only if $\mathrm{CJ}(\Lambda)$ is positive. For now, we show one direction:

Proposition 5.5.11. *If Λ is completely positive, then $\mathrm{CJ}(\Lambda)$ is a positive element of $\mathrm{End}(\mathcal{H}_A \otimes \mathcal{H}_B)$.*

Proof. Recall that it is sufficient to show that $\mathrm{CJ}(\Lambda)$ is positive when $\Lambda \otimes \mathrm{Id}_{\mathrm{End}(\mathcal{H}_A)}$ is positive.

Let $|v\rangle := \sum_i |ii\rangle \in \mathcal{H}_A \otimes \mathcal{H}_A$. By hypothesis $(\Lambda \otimes \mathrm{Id}_{\mathrm{End}(\mathcal{H}_A)})(|v\rangle\langle v|) \in \mathrm{End}(\mathcal{H}_B \otimes \mathcal{H}_A)$ is positive because $|v\rangle\langle v|$ is positive.

We claim

$$\mathrm{CJ}(\Lambda) = (\Lambda \otimes \mathrm{Id}_{\mathrm{End}(\mathcal{H}_A)})(|v\rangle\langle v|),$$

which will prove the proposition. To prove the claim:

$$\begin{aligned}&\Lambda \otimes \mathrm{Id}_{\mathrm{End}(\mathcal{H}_A)}(|v\rangle\langle v|)\\ &= \left[\left(\sum_{ijst} \Lambda^{\binom{i}{j}}_{\binom{s}{t}} |t\rangle\langle s| \otimes (\langle i| \otimes |j\rangle)\right) \otimes \left(\sum_{i'j'} |i'\rangle\langle j'| \otimes (\langle i'| \otimes |j'\rangle)\right)\right]\left(\sum_{kk'} |kk\rangle\langle k'k'|\right).\end{aligned}$$

In the evaluation on $|v\rangle\langle v|$, $\langle i| \otimes \langle i'|$ is contracted with $|kk\rangle$, so one gets zero unless $i = i'$ and similarly $|j\rangle \otimes |j'\rangle$ is contracted with $\langle k'k'|$. Hence

$$\Lambda \otimes \mathrm{Id}_{\mathrm{End}(\mathcal{H}_A)}(|v\rangle\langle v|) = \sum_{ijst} \Lambda^{\binom{i}{j}}_{\binom{s}{t}} |it\rangle\langle js| = \mathrm{CJ}(\Lambda).$$ □

5.5.3. Stinespring and Kraus representations.

Proposition 5.5.12 (Kraus representation [**Cho75**]). *Let $\Lambda : \mathrm{End}(\mathcal{H}_A) \to \mathrm{End}(\mathcal{H}_B)$ be such that $\mathrm{CJ}(\Lambda)$ is positive. Then there exists a collection $Y_\alpha \in \mathrm{Hom}(\mathcal{H}_A, \mathcal{H}_B)$ such that $\Lambda(X) = \sum_\alpha Y_\alpha X Y_\alpha^\dagger$.*

Such a representation of Λ is called a *Kraus representation*.

Proof. By positivity of $\mathrm{CJ}(\Lambda)$, we may write $\mathrm{CJ}(\Lambda) = \sum_\alpha |v_\alpha\rangle\langle v_\alpha|$ for some $|v_\alpha\rangle \in \mathcal{H}_A \otimes \mathcal{H}_B$. Expand out $|v_\alpha\rangle = \sum_{k,u} v_{\alpha,ku} |k\rangle \otimes |u\rangle$. Then

$$|v_\alpha\rangle\langle v_\alpha| = \sum_{j,k,t,u} v_{\alpha,ku}\overline{v_{\alpha,jt}} |k\rangle|u\rangle \otimes \langle j|\langle t|,$$

and

$$\Lambda = \sum_\alpha \mathrm{CJ}^{-1}(|v_\alpha\rangle\langle v_\alpha|) = \sum_\alpha \sum_{kjut} v_{\alpha,ku}\overline{v_{\alpha,jt}} |u\rangle\langle t| \otimes (\langle k| \otimes |j\rangle).$$

Take $Y_\alpha = \sum_{k,u} \upsilon_{\alpha,ku}|u\rangle\langle k|$. Then

$$\begin{aligned}\Lambda(X) &= \sum_\alpha \mathrm{CJ}^{-1}(|\upsilon_\alpha\rangle\langle\upsilon_\alpha|)(X)\\ &= \sum_\alpha\sum_{kjut} \upsilon_{\alpha,ku}\overline{\upsilon_{\alpha,jt}}X_j^k|u\rangle\langle t|\\ &= \sum_\alpha Y_\alpha X Y_\alpha^\dagger.\end{aligned}$$

□

Exercise 5.5.13. Let $\sigma \in \mathcal{D}(\mathcal{H}_C)$. Give a Kraus representation of the map

$$\begin{aligned}\Lambda : \mathrm{End}(\mathcal{H}_A) &\to \mathrm{End}(\mathcal{H}_A \otimes \mathcal{H}_C)\\ X &\mapsto X\otimes\sigma.\end{aligned}$$

⊚

Proposition 5.5.14 (Stinespring representation [**Sti55**]). *Let $\Lambda : \mathrm{End}(\mathcal{H}_A) \to \mathrm{End}(\mathcal{H}_B)$ admit a Kraus representation. Then there exists an isometric embedding $Z : \mathcal{H}_A \to \mathcal{H}_B\otimes\mathcal{H}_C$ for some Hilbert space $\mathcal{H}_C$, such that*

$$\Lambda(X) = \mathrm{trace}_{\mathcal{H}_C}(ZXZ^\dagger). \tag{5.5.3}$$

The map Z is called a *Stinespring dilation* of Λ and the representation (5.5.3) is called a *Stinespring representation.*

Proof. Say there are r Y_α's in a Kraus representation of Λ. Take $\mathcal{H}_C = \mathbb{C}^r$ and $Z := \sum_\alpha Y_\alpha\otimes|\alpha\rangle$. Then

$$ZXZ^\dagger = \left(\sum_\alpha Y_\alpha\otimes|\alpha\rangle\right)X\left(\sum_\beta Y_\beta^\dagger\otimes\langle\beta|\right).$$

Taking the trace over $\mathcal{H}_C$ one recovers Λ in its Kraus representation. □

Exercise 5.5.15. Show that if Λ admits a Stinespring representation, then it admits a Kraus representation.

We have the following implications:

Λ is completely positive $\Rightarrow$ (Proposition 5.5.11) $\mathrm{CJ}(\Lambda)$ is positive $\Rightarrow$ (Proposition 5.5.12) Λ admits a Kraus representation $\Rightarrow$ (Proposition 5.5.14) Λ admits a Stinespring dilation $\Rightarrow$ (Exercise 5.5.6) Λ is completely positive.

In summary:

Proposition 5.5.16. *The following are equivalent:*

- *Λ is completely positive.*
- *$\mathrm{CJ}(\Lambda)$ is positive.*
- *Λ admits a Stinespring representation.*
- *Λ admits a Kraus representation.*

We next characterize trace preserving in terms of Kraus and Stinespring representations:

Exercise 5.5.17. Show that a completely positive $\Lambda : \mathrm{End}(\mathcal{H}_A) \to \mathrm{End}(\mathcal{H}_B)$ is trace preserving if and only if in a Kraus representation $\sum_\alpha Y_\alpha^\dagger Y_\alpha = \mathrm{Id}_{\mathcal{H}_A}$. ⊚

Exercise 5.5.18. Show that a completely positive $\Lambda : \mathrm{End}(\mathcal{H}_A) \to \mathrm{End}(\mathcal{H}_B)$ is trace preserving if and only if $Z^\dagger Z = \mathrm{Id}_{\mathcal{H}_A}$ in a Stinespring representation.

In summary:

Theorem 5.5.19. *Let* $\Lambda : \mathrm{End}(\mathcal{H}_A) \to \mathrm{End}(\mathcal{H}_B)$. *The following are equivalent:*

(1) Λ *is completely positive.*
(2) $\mathrm{CJ}(\Lambda)$ *is positive.*
(3) Λ *admits a Stinespring representation.*
(4) Λ *admits a Kraus representation.*

When Λ *is completely positive, the following are equivalent:*

(1) Λ *is trace preserving, i.e., a quantum channel.*
(2) $\mathrm{trace}_{\mathcal{H}_B}(\mathrm{CJ}(\Lambda)) = \mathrm{Id}_{\mathcal{H}_A}$.
(3) $Z^\dagger Z = \mathrm{Id}_{\mathcal{H}_A}$ *in a Stinespring representation of* Λ.
(4) $\sum_\alpha Y_\alpha^\dagger Y_\alpha = \mathrm{Id}_{\mathcal{H}_A}$ *in a Kraus representation of* Λ.

Exercise 5.5.20. Compute the Stinespring and Kraus representations of $\Lambda : \mathrm{End}(\mathcal{H}_A) \to \mathrm{End}(\mathcal{H}_B)$ when $\mathcal{H}_A = \mathcal{H}_B \otimes \mathcal{H}_E$ and $\Lambda = \mathrm{trace}_{\mathcal{H}_E}$.

Exercise 5.5.21. Show that if $\Lambda : \mathrm{End}(\mathcal{H}_A) \to \mathrm{End}(\mathcal{H}_B)$ is given in its Kraus representation, then $\mathrm{CJ}(\Lambda) = \sigma_{(123)} \circ \sum_\alpha Y_\alpha \otimes Y_\alpha^\dagger$.

5.5.4. Examples of quantum channels.

Example 5.5.22 (No peek measurement). If we make a projective measurement of $\rho \in \mathcal{D}(\mathcal{H})$ based on an orthogonal decomposition $\mathcal{H} = \bigoplus_{s=1}^k \mathcal{M}_s$ and do not look at the outcome, we only know ρ is projected into $\mathcal{M}_s$ with the probability of that event occurring, so we obtain

$$\begin{aligned}\rho \mapsto \sum_s \Pr(\mathrm{Proj}_{\mathcal{M}_s} \cdot \rho) \frac{\mathrm{Proj}_{\mathcal{M}_s} \cdot \rho}{\mathrm{trace}(\mathrm{Proj}_{\mathcal{M}_s} \cdot \rho)} \\ = \sum_s \mathrm{Proj}_{\mathcal{M}_s} \rho \, \mathrm{Proj}_{\mathcal{M}_s}^\dagger .\end{aligned}$$

Complete positivity is clear as the second line is a Kraus representation of the channel, and trace preserving holds because $\mathrm{Proj}_{\mathcal{M}_s}^\dagger \mathrm{Proj}_{\mathcal{M}_s} = \mathrm{Proj}_{\mathcal{M}_s}$ and $\sum_s \mathrm{Proj}_{\mathcal{M}_s} = \mathrm{Id}$.

Example 5.5.23 (Noisy channels). Let $\mathcal{H} = \mathbb{C}^2$ and say a channel $\Lambda : \mathrm{End}(\mathbb{C}^2) \to \mathrm{End}(\mathbb{C}^2)$ transmits a density operator correctly with probability $1 - p$ and that it flips 0 and 1 with probability p. We would get the channel (in Kraus representation):

$$\rho \mapsto (1-p)\,\mathrm{Id}_{\mathbb{C}^2}\,\rho\,\mathrm{Id}_{\mathbb{C}^2} + p\begin{pmatrix} 0 & 1 \\ 1 & 0 \end{pmatrix}\rho\begin{pmatrix} 0 & 1 \\ 1 & 0 \end{pmatrix}.$$

In quantum channels additional kinds of errors may arise, e.g., a "phase flip" $|0\rangle \mapsto |0\rangle$, $|1\rangle \mapsto -|1\rangle$, or something even messier, e.g., $|0\rangle \mapsto i|1\rangle$, $|1\rangle \mapsto -i|0\rangle$. Consider the situation where all three types of errors occur, respectively, with probabilities p_f, p_{ph}, p_m, where $p_f + p_{ph} + p_m = p$. Then our channel would be

$$\begin{aligned}\rho \mapsto &(1-p)\,\mathrm{Id}_{\mathbb{C}^2}\,\rho\,\mathrm{Id}_{\mathbb{C}^2} + p_f\begin{pmatrix} 0 & 1 \\ 1 & 0 \end{pmatrix}\rho\begin{pmatrix} 0 & 1 \\ 1 & 0 \end{pmatrix} \\ &+ p_{ph}\begin{pmatrix} 1 & 0 \\ 0 & -1 \end{pmatrix}\rho\begin{pmatrix} 1 & 0 \\ 0 & -1 \end{pmatrix} + p_m\begin{pmatrix} 0 & -i \\ i & 0 \end{pmatrix}\rho\begin{pmatrix} 0 & -i \\ i & 0 \end{pmatrix}.\end{aligned}$$

Example 5.5.24 (Dephasing channel). Say one is using a quantum channel to transmit classical information. A simple way to get a classical state from a quantum state is to kill the off-diagonal entries. Thus for $X = \sum_{ij} X^i_j |i\rangle\langle j|$ one defines

$$\Lambda(X) := \sum_i X^i_i |i\rangle\langle i|.$$

Example 5.5.25 (Mixed unitary (LOCC) channel). Let $U_s \in \mathbf{U}(\mathcal{H})$ be unitary, $1 \le s \le k$, and let p be a probability distribution on $\{1,\dots,k\}$ for some finite k. Consider the channel

$$\Lambda(\rho) := \sum_{s=1}^{k} p_s U_s \rho U_s^{-1}. \tag{5.5.4}$$

This channel will arise when we study a generalization of teleportation called LOCC; see §6.5.

Chapter 6

Quantum information

Shannon's theory of information is so visionary that its adaptation to the quantum setting is straightforward. Nonetheless, the postulates of quantum mechanics give rise to strange and counterintuitive results when one makes the adaptation. We present a quantum version of Shannon's noiseless channel theorem in §6.1. This requires the introduction of von Neumann entropy. Properties of von Neumann entropy are discussed in §6.2. In §6.3, we introduce conditional von Neumann entropy, state and prove the strong subadditivity of von Neumann entropy, and give variants of it. In §6.4 we discuss information-theoretic consequences of strong subadditivity, including a bound on classical communication using a quantum channel. In §6.5 a generalization of quantum teleportation, called LOCC, is introduced. LOCC has connections with classical results of Hardy-Littlewood-Polya and Birkoff. These classical results and Nielson's theorem on LOCC are also presented. In §6.6 we give an overview of further measures of entanglement.

6.1. The quantum noiseless channel theorem

6.1.1. Set-up for the quantum noiseless channel theorem. In analogy with the classical situation, we have a noiseless communication channel, i.e., an isometry $\Lambda : \mathrm{End}(\mathcal{H}_A) \to \mathrm{End}(\mathcal{H}_B)$. What we are really interested in is $\Lambda^{\otimes d} : \mathrm{End}(\mathcal{H}_A^{\otimes d}) \to \mathrm{End}(\mathcal{H}_B^{\otimes d})$ as $d \to \infty$. Assume $\mathcal{H}_A = \mathcal{H}_B = \mathcal{H}$ and $\Lambda = \mathrm{Id}_{\mathcal{H}}$. We want to apply an encoding channel $\mathcal{E}^d : \mathrm{End}(\mathcal{H}^{\otimes d}) \to \mathcal{H}(d) = (\mathbb{C}^2)^{\otimes dR}$ to compress messages and a decoding channel $\mathcal{D}^d : \mathcal{H}(d) \to \mathrm{End}(\mathcal{H}^{\otimes d})$ that will restore the messages in such a way that transmission is reliable. The picture is

$$\mathcal{H}^{\otimes d} \xrightarrow{\mathcal{E}} \mathcal{H}(d) = (\mathbb{C}^2)^{\otimes dR} \xrightarrow{\mathcal{D}} \mathcal{H}^{\otimes d}$$

and the goal is to have R as small as possible while having $F(\rho^{\otimes d}, \mathcal{E}\circ\mathcal{D}(\rho^{\otimes d})) \to 1$ as $d \to \infty$. To determine R, we need a quantum version of entropy.

6.1.2. von Neumann entropy. Given a density operator ρ, define the operator $\log(\rho)$ as follows: write ρ in terms of its eigenvectors and nonzero eigenvalues, $\rho = \sum_j \lambda_j |\psi_j\rangle\langle\psi_j|$; then $\log(\rho) := \sum_j \log(\lambda_j)|\psi_j\rangle\langle\psi_j|$.

Definition 6.1.1. The *von Neumann entropy* of a density operator ρ is $H(\rho) := -\operatorname{trace}(\rho\log(\rho))$.

If $\rho = \sum_j \lambda_j|\psi_j\rangle\langle\psi_j|$, then $H(\rho) = -\sum_j \lambda_j \log(\lambda_j)$ so if ρ is classical (i.e., diagonal), one obtains the Shannon entropy.

Proposition 6.1.2. *The von Neumann entropy of $\rho \in \mathcal{D}(\mathcal{H})$ has the following properties:*

(1) *$H(\rho) \geq 0$ with equality if and only if ρ is pure.*
(2) *Write $n = \dim\mathcal{H}$. Then $H(\rho) \leq \log(n)$ with equality if and only if $\rho = \frac{1}{n}\operatorname{Id}_{\mathcal{H}}$.*
(3) *If $\rho = |\psi\rangle\langle\psi| \in \operatorname{End}(\mathcal{H}_A \otimes \mathcal{H}_B)$, then $H(\rho_A) = H(\rho_B)$, where $\rho_A = \operatorname{trace}_{\mathcal{H}_B}(\rho) \in \operatorname{End}(\mathcal{H}_A)$ and ρ_B is defined similarly.*

Exercise 6.1.3. Prove the first two assertions.

Proof of (3). Using Proposition 5.4.6, write $|\psi\rangle = \sum_j z_j|\phi_j\rangle\otimes|\eta_j\rangle$ where the $|\phi_j\rangle \in \mathcal{H}_A$ are orthonormal as are the $|\eta_j\rangle \in \mathcal{H}_B$ and where $\sum |z_j|^2 = 1$. Then $|\psi\rangle\langle\psi| = \sum_{i,j} z_i\bar{z}_j|\phi_i\rangle|\eta_i\rangle\otimes\langle\phi_j|\langle\eta_j|$, so

$$\operatorname{trace}_{\mathcal{H}_B}(|\psi\rangle\langle\psi|) = \sum_j |z_j|^2|\phi_j\rangle\langle\phi_j| \quad \text{and}$$
$$\operatorname{trace}_{\mathcal{H}_A}(|\psi\rangle\langle\psi|) = \sum_j |z_j|^2|\eta_j\rangle\langle\eta_j|.$$

Since the nonzero eigenvalues are the same, the entropies will be the same. □

Contrast (3) with the classical situation, where

$$H(\bar{p}_{\mathcal{X}\times\mathcal{Y}}) \geq \max\{H(\bar{p}_{\mathcal{X}}), H(\bar{p}_{\mathcal{Y}})\}$$

where $\bar{p}_{\mathcal{X}}, \bar{p}_{\mathcal{Y}}$ are the marginal distributions. In the situation of (3), given $|\psi\rangle \in \mathcal{H}_A\otimes\mathcal{H}_B$, $H(\rho_{AB}) = H(|\psi\rangle\langle\psi|) = 0$ and $H(\rho_A) = H(\rho_B) \geq 0$ with strict inequality occurring whenever, considered as a bilinear form, $\operatorname{rank}(|\psi\rangle) > 1$.

6.1.3. Quantum typical subspace theorem. Recall that in the classical case, Theorem 4.4.1, a probability distribution on $\mathcal{X}$ induces a splitting $\mathcal{X}^d = \mathcal{X}^d_{\epsilon-\text{typ}} \sqcup \mathcal{X}^d_{\delta-\text{small}}$ (Theorem 4.4.1). Here is a quantum analog:

Theorem 6.1.4 (Quantum typical subspace theorem). *Given $(\mathcal{H}, \rho)$ and $\epsilon, \delta > 0$, there exists d_0 such that for all $d \geq d_0$, there exists an orthogonal direct sum decomposition $\mathcal{H}^{\otimes d} = \mathcal{H}^d_{\epsilon-\text{typ}} \oplus \mathcal{H}^d_{\delta-\text{small}}$ satisfying:*

(1) $\Pr(\mathcal{H}^d_{\delta-\text{small}}) < \delta$; *i.e.,* $\operatorname{trace}(\operatorname{Proj}_{\mathcal{H}^d_{\delta-\text{small}}} \rho^{\otimes d}) < \delta$.

(2) $(1-\delta)2^{d(H(\rho)-\epsilon)} \leq \dim(\mathcal{H}^d_{\epsilon-\text{typ}}) \leq 2^{d(H(\rho)+\epsilon)}$.

Proof. Assume $\operatorname{rank}(\rho) = n$. Write $\rho = \sum_{j=1}^n \lambda_j |x_j\rangle\langle x_j|$, where the $|x_j\rangle$ are orthonormal. Observe that $\rho^{\otimes d}$ has a basis of orthonormal eigenvectors consisting of tensor products of eigenvectors of ρ. Write $|\bar{x}\rangle = |x_{j_1} \cdots x_{j_d}\rangle$, which is an eigenvector for $\rho^{\otimes d}$ with eigenvalue $\lambda(\bar{x}) := \lambda_1^{c_1} \cdots \lambda_n^{c_n}$, where c_j is the number of times x_j appears in $\bar{x}$, so $c_1 + \cdots + c_n = d$. As in the classical case, set $\operatorname{wt}_d(\bar{x}^d) = (c_1, \ldots, c_n)$, so $E[\operatorname{wt}_d] = (d\lambda_1, \ldots, d\lambda_n)$. Take

$$\mathcal{H}^d_{\epsilon-\text{typ}} = \operatorname{span}\left\{\text{eigenvectors } |\bar{x}^d\rangle \,\middle|\, \left\|\frac{1}{d}(\operatorname{wt}_d(\bar{x}^d) - E[\operatorname{wt}_d])\right\|_1 < \frac{\epsilon}{\log(n)}\right\}.$$

Now argue as in the classical case. □

Remark 6.1.5. A geometric description of $\mathcal{H}^d_{\epsilon-\text{typ}}$ in terms of the representation theory of $\mathbf{U}(\mathcal{H}) \times \mathfrak{S}_d$, where $\mathfrak{S}_d$ denotes the permutation group, and a proof of its existence without utilizing the law of large numbers will be given in Theorem 7.2.6.

Here is a quantum analog of Theorem 4.4.3:

Theorem 6.1.6. *Given $(\mathcal{H}, \rho)$, let $R < H(\rho)$ and $\eta > 0$. Then there exists d_0 such that for all $d \geq d_0$ and all subspaces $\mathcal{M}(d) \subset \mathcal{H}^{\otimes d}$ with $\dim \mathcal{M}(d) \leq 2^{dR}$,*

$$\operatorname{trace}(\operatorname{Proj}_{\mathcal{M}(d)} \rho^{\otimes d}) \leq \eta.$$

Proof. Write

$$\rho^{\otimes d} = \sum_{\substack{\text{eigenvectors} \\ \bar{x} \in \mathcal{H}^d_{\epsilon-\text{typ}}}} \lambda(\bar{x})|\bar{x}\rangle\langle\bar{x}| + \sum_{\substack{\text{eigenvectors} \\ \bar{z} \in \mathcal{H}^d_{\delta-\text{small}}}} \lambda(\bar{z})|\bar{z}\rangle\langle\bar{z}|.$$

Then

$$\operatorname{trace}(\operatorname{Proj}_{\mathcal{M}(d)} \rho^{\otimes d}) \leq \sum_{\substack{\text{eigenvectors} \\ \bar{x} \in \mathcal{H}^d_{\epsilon-\text{typ}}}} \operatorname{trace}\Big(\operatorname{Proj}_{\mathcal{M}(d)} \lambda(\bar{x})|\bar{x}\rangle\langle\bar{x}|\Big) + \delta.$$

Since $\lambda(\bar{x}) \leq 2^{-dH(\rho)+\epsilon}$ and $\dim \mathcal{M}(d) \leq 2^{dR}$, the first term is at most

$$2^{d(R-H(\rho)+\epsilon)}.$$

Recalling that $R-H(\rho)$ is negative, choose ϵ sufficiently small that $R-H(\rho)+\epsilon < 0$ and choose δ, d_0 such that $2^{d_0(R-H(\rho)+\epsilon)} + \delta < \eta$. □

6.1.4. The quantum noiseless channel theorem.

Theorem 6.1.7 ([**Sch95**]). *Let $(\mathcal{H}, \rho)$ be a quantum source. For all $R > H(\rho)$, there exist compression spaces $\mathcal{H}(d)$ of dimension 2^{dR}, encoder channels $\mathcal{E}^d : \mathrm{End}(\mathcal{H}^{\otimes d}) \to \mathrm{End}(\mathcal{H}(d))$, and decoder channels $\mathcal{D}^d : \mathrm{End}(\mathcal{H}(d)) \to \mathrm{End}(\mathcal{H}^{\otimes d})$ such that*

$$\lim_{d\to\infty} F(\rho^{\otimes d}, \mathcal{D}^d \circ \mathcal{E}^d(\rho^{\otimes d})) = 1.$$

If $R < H(\rho)$, then any compression scheme is unreliable.

The idea of the proof is the same as the classical case. Here are the details.

Proof. Say $R > H(\rho)$ and write $R = H(\rho) + \epsilon$. The quantum typical subspace theorem implies that for all $\delta > 0$ and d sufficiently large, there exists $\mathcal{H}^d_{\epsilon-\mathrm{typ}} \subset \mathcal{H}^{\otimes d}$ such that

$$\mathrm{trace}(\rho^{\otimes d}\,\mathrm{Proj}_{\mathcal{H}^d_{\epsilon-\mathrm{typ}}}) \geq 1-\delta \quad \text{and} \quad \dim(\mathcal{H}^d_{\epsilon-\mathrm{typ}}) \leq 2^{dR}.$$

Fix $\mathcal{H}(d) = \mathbb{C}^{2^{dR}}$ and an isometric inclusion $\mathcal{H}^d_{\epsilon-\mathrm{typ}} \subseteq \mathcal{H}(d)$ that we will suppress reference to in the remainder of the proof. Let $P_{0,d} : \mathcal{H}^{\otimes d} \to \mathbb{C}|\psi_d\rangle$ be an operator sending $\mathcal{H}^d_{\epsilon-\mathrm{typ}}$ to zero and $\mathcal{H}^d_{\delta-\mathrm{small}}$ to $\mathbb{C}|\psi_d\rangle$, where $|\psi_d\rangle \in \mathcal{H}(d)$ is any vector. Define the encoder channels

$$\begin{aligned}\mathcal{E}^d &: \mathcal{H}^{\otimes d} \to \mathcal{H}(d)\\ \rho^{\otimes d} &\mapsto \mathrm{Proj}_{\mathcal{H}^d_{\epsilon-\mathrm{typ}}}\rho^{\otimes d}\,\mathrm{Proj}_{\mathcal{H}^d_{\epsilon-\mathrm{typ}}} + P_{0,d}\rho^{\otimes d}P_{0,d},\end{aligned}$$

and the decoder channels $\mathcal{D}^d : \mathcal{H}(d) \to \mathcal{H}^{\otimes d}$ to be the orthogonal projection onto $\mathcal{H}^d_{\epsilon-\mathrm{typ}}$. We estimate the fidelity:

$$\begin{aligned}
&F(\rho^{\otimes d}, \mathcal{D}^d \circ \mathcal{E}^d(\rho^{\otimes d}))\\
&= \left[\mathrm{trace}\left(\left(\sqrt{\rho^{\otimes d}}\mathcal{D}^d \circ \mathcal{E}^d\rho^{\otimes d}\sqrt{\rho^{\otimes d}}\right)^{\frac{1}{2}}\right)\right]^2 \quad \text{by Exercise 5.3.6}\\
&= \left[\mathrm{trace}\left(\left(\sqrt{\rho^{\otimes d}}\,\mathrm{Proj}_{\mathcal{H}^d_{\epsilon-\mathrm{typ}}}\rho^{\otimes d}\,\mathrm{Proj}_{\mathcal{H}^d_{\epsilon-\mathrm{typ}}}\sqrt{\rho^{\otimes d}} + \sqrt{\rho^{\otimes d}}P_{0,d}\rho^{\otimes d}P_{0,d}^{\dagger}\sqrt{\rho^{\otimes d}}\right)^{\frac{1}{2}}\right)\right]^2\\
&\geq \left[\mathrm{trace}\left(\left(\sqrt{\rho^{\otimes d}}\,\mathrm{Proj}_{\mathcal{H}^d_{\epsilon-\mathrm{typ}}}\rho^{\otimes d}\,\mathrm{Proj}_{\mathcal{H}^d_{\epsilon-\mathrm{typ}}}\sqrt{\rho^{\otimes d}}\right)^{\frac{1}{2}}\right)\right]^2\\
&= [\mathrm{trace}(\rho^{\otimes d}\,\mathrm{Proj}_{\mathcal{H}^d_{\epsilon-\mathrm{typ}}})]^2 \quad \text{because } \mathrm{Proj}_{\mathcal{H}^d_{\epsilon-\mathrm{typ}}}, \sqrt{\rho^{\otimes d}} \text{ commute}\\
&\geq (1-\delta)^2.
\end{aligned}$$

Now just let $\delta \to 0$.

The proof of the other direction is similar to the classical case; see Exercise 6.1.8. □

Exercise 6.1.8. Prove that for any $\mathcal{D}^d, \mathcal{E}^d$, when $\dim \mathcal{H}(d) < 2^{dH(\rho)}$, then $F(\rho^{\otimes d}, \mathcal{D}^d \circ \mathcal{E}^d(\rho^{\otimes d}))$ becomes arbitrarily small as $d \to \infty$. ⊚

6.1.5. Other quantum channel capacities. A natural next step would be to prove a noisy quantum channel theorem. Such a theorem is beyond the scope of this book. See [**Wat18**, Thm. 8.55] for a theorem in this direction.

In addition to conveying quantum states, quantum channels may be used to convey:

- classical messages,
- classical messages assisted with entanglement, as in super-dense coding, §2.4,
- quantum states assisted with a classical channel and entanglement, as with teleportation, §2.5.

We will investigate these for noiseless quantum channels. Before doing so, we need a better understanding of von Neumann entropy and its variants.

6.2. Properties of von Neumann entropy

We prove several inequalities regarding von Neumann entropy, summarized in §6.2.6. The main tool for proving the inequalities is the quantum relative von Neumann entropy introduced in §6.2.2. Before discussing it, we introduce its classical cousin in §6.2.1. The key property both relative entropies share is that they are nonnegative. Relative entropy will be used primarily as a tool for comparing probability distributions in the classical case and density operators in the quantum case. In detail, §6.2.3, §6.2.4, §6.2.5, respectively, cover von Neumann entropy and projective measurements, composite systems, and classical/quantum systems.

6.2.1. Relative entropy: Classical case.

Definition 6.2.1. Given two probability distributions $\bar{p}, \bar{q}$ on a finite set $\mathcal{X} = [n]$, with $\text{supp}(\bar{p}) \subseteq \text{supp}(\bar{q})$, the *relative entropy*, also called the *Kullback-Leibler divergence*, is

$$\begin{aligned} H(\bar{p}||\bar{q}) :&= -\sum_{i=1}^{n} p_i \log \frac{q_i}{p_i} \\ &= -H(\bar{p}) - \sum_i p_i \log(q_i). \end{aligned}$$

If $\text{supp}(\bar{p}) \not\subseteq \text{supp}(\bar{q})$, one takes $H(\bar{p}||\bar{q}) = \infty$. This convention is used to make relative entropy continuous.

Exercise 6.2.2. Show that $H(\bar{p}||\bar{q}) = E_{\bar{p}}[\log \frac{p(x)}{q(x)}]$.

Exercise 6.2.3. For a probability distribution $p_{\mathcal{X}\times\mathcal{Y}}$ on $\mathcal{X}\times\mathcal{Y}$, show that

$$I(p_{\mathcal{X}} : p_{\mathcal{Y}}) = H(p_{\mathcal{X}\times\mathcal{Y}}||p_{\mathcal{X}} \times p_{\mathcal{Y}}),$$

where $p_{\mathcal{X}}, p_{\mathcal{Y}}$ are the marginal distributions of $p_{\mathcal{X}\times\mathcal{Y}}$ and I is the mutual information (see Definition 4.5.6).

Note that the relative entropy is zero when $\bar{p} = \bar{q}$. We first observe it is always nonnegative:

Proposition 6.2.4. $H(\bar{p}||\bar{q}) \geq 0$.

Proof. Observe that $\log(x)\ln(2) = \ln x \leq x - 1$ for all $x > 0$ with equality if and only if $x = 1$; i.e., $\log(x) \leq \frac{1}{\ln(2)}(x-1)$. We have

$$\begin{aligned} H(\bar{p}||\bar{q}) &= -\sum_{i=1}^{n} p_i \log \frac{q_i}{p_i} \\ &\geq -\sum_i p_i \frac{1}{\ln(2)}\left(\frac{q_i}{p_i} - 1\right) \\ &= \frac{1}{\ln(2)}\sum_i (p_i - q_i) \\ &= \frac{1}{\ln(2)}(1-1). \end{aligned}$$

□

We next quantify the positivity:

Lemma 6.2.5 (Pinsker's inequality)**.** *Let $\bar{p}, \bar{q}$ be probability distributions on $[n]$. Then*

$$H(\bar{p}||\bar{q}) \geq \frac{1}{2\ln 2}||\bar{p} - \bar{q}||_1^2.$$

Proof. First consider the special case $n = 2$, so $\bar{p} = (p, 1-p)$, $\bar{q} = (q, 1-q)$. Assume $p \geq q$ (otherwise exchange p with $1-p$ and q with $1-q$).

Consider

$$\begin{aligned} f(p,q) &:= H(\bar{p}||\bar{q}) - \frac{1}{2\ln 2}||\bar{p} - \bar{q}||_1^2 \\ &= p\log\frac{p}{q} + (1-p)\log\frac{1-p}{1-q} - \frac{1}{2\ln 2}(2(p-q))^2. \end{aligned}$$

Note that $f(p,q) = 0$ for $q = p$ and we need to show $f(p,q) \geq 0$ for $q < p$.

Exercise 6.2.6. Show that $\frac{\partial f}{\partial q} \leq 0$, so if we decrease q, then f increases, to prove the special case.

Now for the general case, set $A := \{j \in [n] \mid p_j \geq q_j\}$ and write $p_A = (\sum_{j\in A} p_j, \sum_{j\notin A} p_j)$ and $q_A = (\sum_{j\in A} q_j, \sum_{j\notin A} q_j)$. Then

$$\begin{aligned}
\|\bar{p}-\bar{q}\|_1 &= \sum_j |p_j - q_j| \\
&= \sum_{j\in A} p_j - q_j + \sum_{j\notin A} q_j - p_j \\
&= \left|\sum_{j\in A} p_j - q_j\right| + \left|\left(1-\sum_{j\notin A} p_j\right) - \left(1-\sum_{j\notin A} q_j\right)\right| \\
&= \|p_A - q_A\|_1.
\end{aligned}$$

Let Z be the random variable taking 1 on A and 0 on $[n]\backslash A$. Let $H(\bar{p}||\bar{q}|Z)$ be the relative entropy conditioned on Z. Then

$$H(\bar{p}||\bar{q}) = H(p_Z||q_Z) + H(\bar{p}||\bar{q}|Z).$$

By Proposition 6.2.4, all three of these quantities are nonnegative. Observe that $H(p_Z||q_Z) = H(p_A||q_A)$. Putting it all together

$$H(\bar{p}||\bar{q}) \geq H(p_A||q_A) \geq \frac{1}{2\ln 2}\|p_A - q_A\|_1^2 = \frac{1}{2\ln 2}\|p-q\|_1^2. \qquad \square$$

One interpretation of relative entropy is as follows: say that $\mathcal{X}$ sends symbols according to a distribution p, but the encoder thinks symbols are being sent according to a distribution q and compresses accordingly. Then one needs the compression space $\mathcal{Y}(d)$ of size about $2^{H(\bar{p})+H(\bar{p}||\bar{q})}$ to ensure reliable transmission.

6.2.2. Relative entropy: Quantum case.

Definition 6.2.7. Let $\rho, \sigma \in \mathcal{D}(\mathcal{H})$ with $\text{Image}(\rho) \subseteq \text{Image}(\sigma)$. The *relative von Neumann entropy* is

$$H(\rho||\sigma) := \text{trace}(\rho\log(\rho)) - \text{trace}(\rho\log(\sigma)).$$

When $\text{Image}(\rho) \nsubseteq \text{Image}(\sigma)$, one takes $H(\rho||\sigma) = \infty$ to be consistent with continuity.

The relative von Neumann entropy shares the positivity property of its classical cousin:

Proposition 6.2.8 ([Kle31]). *For density operators ρ, σ on a Hilbert space $\mathcal{H}$, $H(\rho||\sigma) \geq 0$ with equality if and only if $\rho = \sigma$.*

Proof. Assume ρ, σ are both of full rank. Write the eigenvalue decompositions $\rho = \sum_i p_i |v_i\rangle\langle v_i|$, $\sigma = \sum_j q_j |w_j\rangle\langle w_j|$. Then

$$\begin{aligned}
H(\rho||\sigma) &= \sum_i p_i \log(p_i) - \operatorname{trace}\left(\left(\sum_i p_i |v_i\rangle\langle v_i|\right)\left(\sum_j \log(q_j)|w_j\rangle\langle w_j|\right)\right) \\
&= \sum_i p_i \log(p_i) - \operatorname{trace}\left(\left[\sum_{i,j} p_i \log(q_j)\langle v_i|w_j\rangle\right]|v_i\rangle\langle w_j|\right) \\
&= \sum_i p_i \log(p_i) - \sum_{i,j} p_i \log(q_j)\langle v_i|w_j\rangle\langle w_j|v_i\rangle.
\end{aligned}$$

Write $P_{ij} = \langle v_i|w_j\rangle\langle w_j|v_i\rangle$. Note that $P_{ij} \geq 0$ for all i, j, and for all j, $\sum_i P_{ij} = |w_j|^2 = 1$, and for all i, $\sum_j P_{ij} = 1$. Thus the matrix (P_{ij}) is doubly stochastic (cf. §1.6).

Recall $\log(x)$ is a concave function (cf. §4.3.1), so

$$\sum_j P_{ij} \log(q_j) \leq \log\left(\sum_j P_{ij} q_j\right).$$

Set $\tilde{q}_i = \sum_j P_{ij} q_j$ and recall from §1.6 that $\tilde{q}_i$ is a probability distribution. Thus $H(\rho||\sigma) \geq \sum_i p_i \log(\frac{\tilde{q}_i}{p_i}) = H(\bar{p}||\bar{\tilde{q}}) \geq 0$.

For the case ρ, σ are not of full rank, one argues by continuity. □

6.2.3. von Neumann entropy and projective measurements.

Proposition 6.2.9 (von Neumann entropy is nondecreasing under projective measurements). *Let $\mathcal{H} = \oplus_{s=1}^m \mathcal{M}_s$ be an orthogonal direct sum decomposition and write* $\operatorname{Proj}_s : \mathcal{H} \to \mathcal{M}_s$ *for the corresponding orthogonal projection operators. Set $\rho' = \sum_s \operatorname{Proj}_s \rho \operatorname{Proj}_s$. Then $H(\rho') \geq H(\rho)$ with equality if and only if $\rho' = \rho$.*

Exercise 6.2.10. Let $\rho = |\psi\rangle\langle\psi|$ with $\psi = \cos(\theta)|1\rangle + \sin(\theta)|2\rangle$ and let Proj_s be projection onto $\mathbb{C}|s\rangle$ for $s = 1, 2$. For which values of θ does $\rho = \rho'$? Verify that these are the only cases where $H(\rho') = H(\rho)$.

Proof. We have $0 \leq H(\rho||\rho') = -H(\rho) - \operatorname{trace}(\rho \log(\rho'))$. Now since $\sum_s \operatorname{Proj}_s = \operatorname{Id}$,

$$\begin{aligned}
\operatorname{trace}(\rho \log(\rho')) &= \operatorname{trace}\left(\sum_s \operatorname{Proj}_s \rho \log(\rho')\right) \\
&= \operatorname{trace}\left(\sum_s \operatorname{Proj}_s \rho \log(\rho') \operatorname{Proj}_s\right)
\end{aligned}$$

because $\mathrm{Proj}_s^2 = \mathrm{Proj}_s$ and $\mathrm{trace}(ABC) = \mathrm{trace}(BCA)$. Using that Proj_s commutes with ρ' and $\log(\rho')$ because $\mathrm{Proj}_s \mathrm{Proj}_t = 0$ if $s \neq t$, we obtain

$$\begin{aligned}\mathrm{trace}(\rho \log(\rho')) &= \mathrm{trace}\left(\sum_s \mathrm{Proj}_s \rho \, \mathrm{Proj}_s \log(\rho')\right)\\ &= \mathrm{trace}(\rho' \log(\rho'))\\ &= -H(\rho').\end{aligned}$$ □

6.2.4. von Neumann entropy and composite systems. For the classical noisy channel theorem we needed to study the entropy of joint probability distributions and their marginal distributions. In the quantum situation there will be additional motivation for studying such entropic relations when we examine entanglement as a resource, generalizing our examples from Chapter 2. In what follows ρ_{AB} will be a density operator on $\mathcal{H}_A \otimes \mathcal{H}_B$ and $\rho_A = \mathrm{trace}_{\mathcal{H}_B}(\rho_{AB})$, $\rho_B = \mathrm{trace}_{\mathcal{H}_A}(\rho_{AB})$ are, respectively, the induced marginal density operators on $\mathcal{H}_A, \mathcal{H}_B$.

Definition 6.2.11. When $\rho_{AB} = \rho_A \otimes \rho_B$, we say ρ_{AB} is a *product state*.

Exercise 6.2.12. Assume $\rho \in \mathcal{D}(\mathcal{H}_A), \sigma \in \mathcal{D}(\mathcal{H}_B)$ are full rank. Show that

$$\log(\rho \otimes \sigma) = \log(\rho) \otimes \mathrm{Id}_{\mathcal{H}_B} + \mathrm{Id}_{\mathcal{H}_A} \otimes \log(\sigma).$$

More generally show that

$$\log(\rho \otimes \sigma) = \log(\rho) \otimes \mathrm{Id}_{\mathcal{H}'_B} + \mathrm{Id}_{\mathcal{H}'_A} \otimes \log(\sigma),$$

where $\mathcal{H}'_A, \mathcal{H}'_B$ are the spaces, respectively, orthogonal to $\ker(\rho)$, $\ker(\sigma)$. ⊚

Recall that Shannon entropy is subadditive; see (4.3.2). The same holds for von Neumann entropy:

Proposition 6.2.13 (Subadditivity of von Neumann entropy, [**DM35**]). *Let $\rho_{AB} \in \mathcal{D}(\mathcal{H}_A \otimes \mathcal{H}_B)$. Then*

$$H(\rho_{AB}) \leq H(\rho_A) + H(\rho_B)$$

with equality if and only if ρ_{AB} is a product state.

Proof. Assume for now that ρ_A, ρ_B are of full rank.

$$\begin{aligned}
0 &\leq H(\rho_{AB}||\rho_A \otimes \rho_B) \\
&= -H(\rho_{AB}) - \operatorname{trace}(\rho_{AB} \log(\rho_A \otimes \rho_B)) \\
&= -H(\rho_{AB}) - \operatorname{trace}(\rho_{AB}(\log(\rho_A) \otimes \operatorname{Id}_{\mathcal{H}_B})) - \operatorname{trace}(\rho_{AB}(\operatorname{Id}_{\mathcal{H}_A} \otimes \log(\rho_B))) \\
&= -H(\rho_{AB}) - \operatorname{trace}_{\mathcal{H}_A} \operatorname{trace}_{\mathcal{H}_B}(\rho_{AB} \log(\rho_A) \otimes \operatorname{Id}_{\mathcal{H}_B}) \\
&\quad - \operatorname{trace}_{\mathcal{H}_B} \operatorname{trace}_{\mathcal{H}_A}(\rho_{AB} \log(\operatorname{Id}_{\mathcal{H}_A} \otimes \rho_B)) \\
&= -H(\rho_{AB}) - \operatorname{trace}_{\mathcal{H}_A}(\rho_A \log(\rho_A)) - \operatorname{trace}_{\mathcal{H}_B}(\rho_B \log(\rho_B)) \\
&= -H(\rho_{AB}) + H(\rho_A) + H(\rho_B).
\end{aligned}$$

Exercise 6.2.14. Prove the general case where ρ_A, ρ_B need not have full rank. □

Recall that $H(p_{\mathcal{X}}) \leq H(p_{\mathcal{X}\times\mathcal{Y}})$, but this inequality may fail for quantum states. The following result partially salvages the inequality:

Proposition 6.2.15 (Triangle inequality for von Neumann entropy). *Let $\rho_{AB} \in \mathcal{D}(\mathcal{H}_A \otimes \mathcal{H}_B)$. Then*

$$H(\rho_{AB}) \geq |H(\rho_A) - H(\rho_B)|.$$

Proof. Let $\mathcal{H}_R \simeq (\mathcal{H}_A \otimes \mathcal{H}_B)^*$ be a space purifying ρ_{AB}; i.e.,

$$\rho_{ABR} \in \mathcal{D}(\mathcal{H}_A \otimes \mathcal{H}_B \otimes \mathcal{H}_R)$$

is a pure state such that $\operatorname{trace}_{\mathcal{H}_R}(\rho_{ABR}) = \rho_{AB}$. Subadditivity implies $H(\rho_R) + H(\rho_A) \geq H(\rho_{AR})$, and since ρ_{ABR} is pure we have $H(\rho_R) = H(\rho_{AB})$ and $H(\rho_{AR}) = H(\rho_B)$. Putting these together gives $H(\rho_{AB}) \geq H(\rho_B) - H(\rho_A)$ and combining this with the analogous calculation with the roles of A and B reversed gives the result. □

6.2.5. Von Neumann entropy of classical/quantum systems.

When a quantum channel is used to transmit classical information, there is a classical source $\mathcal{X}$ transmitting symbols according to some probability distribution $\bar{p}$ on $\mathcal{X}$. When $j \in \mathcal{X}$ is emitted, Alice transmits a density operator ρ_j to Bob. Bob then applies a measurement channel in an attempt to recover the original classical message sent. To analyze the set-up, as with the no peek measurement Example 5.5.22, we may combine all the density operators into one density operator

$$\rho = \sum_{j \in \mathcal{X}} p_j \rho_j \in \mathcal{D}(\mathcal{H}_A).$$

This section examines entropic properties of such density operators.

Here is a warm-up proposition in the special case where the ρ_j have disjoint orthogonal support:

Proposition 6.2.16. *Let $\mathcal{H} = \bigoplus_{s=1}^m \mathcal{M}_s$ be an orthogonal decomposition. If $\rho = \bigoplus_{s=1}^m p_s\rho_s$ where $\rho_s \in \mathcal{D}(\mathcal{M}_s)$ and $\bar{p}$ is a probability distribution on $[m]$, then $H(\rho) = H(\bar{p}) + \sum_s p_s H(\rho_s)$.*

Proof. Let λ_{s,j_s} be the nonzero eigenvalues of ρ_s.

$$\begin{aligned} H\left(\sum_s p_s\rho_s\right) &= -\sum_s\sum_{j_s}(p_s\lambda_{s,j_s})\log(p_s\lambda_{s,j_s}) \\ &= -\sum_s\sum_{j_s} p_s\lambda_{s,j_s}\log(p_s) - \sum_s\sum_{j_s} p_s\lambda_{s,j_s}\log(\lambda_{s,j_s}) \\ &= H(\bar{p}) + \sum_s p_s H(\rho_s), \end{aligned}$$

where the last line holds because for each fixed s, $\sum_{j_s}\lambda_{s,j_s} = 1$. □

Exercise 6.2.17. Show that if $\rho = \sum_s p_s|s\rangle\langle s|\otimes\rho_s \in \mathcal{D}(\mathbb{C}^m\otimes\mathcal{H})$, then

$$H(\rho) = H(\bar{p}) + \sum_s p_s H(\rho_s). \tag{6.2.1}$$

Recall from Exercise 4.3.5 that in the classical case, entropy cannot decrease under a mixing operation (action of a doubly stochastic matrix). The analogous statement holds in the quantum setting with classical mixing, which generalizes Proposition 6.2.16. As mentioned above, this is the set-up for the quantum transmission of classical messages.

Proposition 6.2.18 (Concavity of von Neumann entropy). *Let ρ_j, $1 \le j \le m$, be density operators on $\mathcal{H}$, let $\bar{p}$ be a probability distribution on $[m]$, and consider the density operator $\sum_j p_j\rho_j \in \mathcal{D}(\mathcal{H})$. Then*

$$H\left(\sum_j p_j\rho_j\right) \ge \sum_j p_j H(\rho_j).$$

In other words, the uncertainty about the mixture of states is at least the weighted average of the uncertainty of the individual states.

Proof. Let $\mathcal{H}_{\mathcal{X}} = \mathbb{C}^m$ and let

$$\rho_{\mathcal{H}\mathcal{X}} := \sum_j p_j\rho_j\otimes|j\rangle\langle j|.$$

Then $\sum_j p_j\rho_j = \rho_{\mathcal{H}} = \text{trace}_{\mathcal{H}_{\mathcal{X}}}\rho_{\mathcal{H}\mathcal{X}}$ and $\rho_{\mathcal{X}} = \sum_j p_j|j\rangle\langle j|$, so $H(\rho_{\mathcal{X}}) = H(\bar{p})$. By Exercise 6.2.17, $H(\rho_{\mathcal{H}\mathcal{X}}) = H(\bar{p}) + \sum_j p_j H(\rho_j)$. On the other hand, by

Proposition 6.2.13, $H(\rho_{\mathcal{HX}}) \leq H(\rho_{\mathcal{H}}) + H(\rho_{\mathcal{X}})$. We have

$$H(\bar{p}) + \sum_j p_j H(\rho_j) \leq H(\rho_{\mathcal{H}}) + H(\rho_{\mathcal{X}}) = H\left(\sum_j p_j \rho_j\right) + H(\bar{p}). \qquad \square$$

We may also upper bound $H(\sum_j p_j \rho_j)$:

Proposition 6.2.19. *Let ρ_j, $1 \leq j \leq m$, be density operators on $\mathcal{H}$, let $\bar{p}$ be a probability distribution on $[m]$, and consider $\sum_j p_j \rho_j \in \mathcal{D}(\mathcal{H})$. Then*

$$H\left(\sum_j p_j \rho_j\right) \leq \sum_j p_j H(\rho_j) + H(\bar{p}),$$

with equality holding if and only if the spaces $\operatorname{supp}(\rho_j)$ *are mutually orthogonal.*

Proposition 6.2.19 may be understood as saying that the uncertainty about $\sum_j p_j \rho_j$ is bounded above by the average of the uncertainties of the ρ_j plus the uncertainty about the index j.

Consider what happens when we try to prove the result naïvely, even in the special case where each ρ_j is pure, $\rho_j = |\psi_j\rangle\langle\psi_j|$. One gets

$$\begin{aligned} H\left(\sum_j p_j \rho_j\right) &= \operatorname{trace}\left(\left(\sum_j p_j |\psi_j\rangle\langle\psi_j|\right)\left(\sum_k \log(p_k)|\psi_k\rangle\langle\psi_k|\right)\right) \\ &= \sum_{j,k} p_j \log(p_k)\langle\psi_j|\psi_k\rangle \operatorname{trace}(|\psi_j\rangle\langle\psi_k|) \\ &= \sum_{j,k} p_j \log(p_k)\langle\psi_j|\psi_k\rangle\langle\psi_k|\psi_j\rangle \\ &= \sum_{j,k} p_j \log(p_k)|\langle\psi_j|\psi_k\rangle|^2, \end{aligned}$$

and we need to show this is at most $\sum_j p_j \log p_j$. One can complete the proof this way, but we instead will take a purification which will simplify the calculation.

Exercise 6.2.20. Finish the proof of the special case by this method. ⊚

Proof. Write $\rho_{\mathcal{H}} = \sum_j p_j \rho_j \in \mathcal{D}(\mathcal{H})$. First consider the case where each ρ_j is pure, $\rho_j = |\psi_j\rangle\langle\psi_j|$. Introduce the purification $\rho_{\mathcal{HX}} := |\phi\rangle\langle\phi| \in \mathcal{D}(\mathcal{H}\otimes\mathbb{C}^m)$ with $|\phi\rangle = \sum_j \sqrt{p_j}|\psi_j\rangle\otimes|j\rangle$. Then

$$H(\rho_{\mathcal{X}}) = H(\rho_{\mathcal{H}}) = H\left(\sum_j p_j |\psi_j\rangle\langle\psi_j|\right).$$

Measure $\rho_{\mathcal{X}}$ in the $|j\rangle$ basis, but do not look at the result. The outcome is $\sum_j p_j|j\rangle\langle j|$. Since entropy is nonincreasing under measurements, $H(\rho_{\mathcal{X}}) \leq H(\sum_j p_j|j\rangle\langle j|) = H(\bar{p})$, proving the special case.

For the general case, write $\rho_j = \sum_{i_j} q_{j,i_j}|e_{j,i_j}\rangle\langle e_{j,i_j}|$. The special case provides the inequality

$$\begin{aligned}
H(\rho_{\mathcal{H}}) &\leq -\sum_{j,i_j} p_j q_{j,i_j}\log(p_j q_{j,i_j}) \\
&= -\sum_{j,i_j} p_j q_{j,i_j}\log(p_j) - \sum_{j,i_j} p_j q_{j,i_j}\log(q_{j,i_j}) \\
&= -\sum_j p_j\log(p_j) - \sum_j p_j\left(\sum_{i_j} q_{j,i_j}\log(q_{j,i_j})\right) \\
&= H(\bar{p}) + \sum_j p_j H(\rho_j).
\end{aligned}$$

The equality case follows from the pure case, recalling that for each fixed j, the e_{j,i_j} are orthonormal. □

6.2.6. Summary of entropic inequalities so far. Von Neumann entropy is nondecreasing under projective measurements: write $\mathcal{H} = \oplus_s \mathcal{M}_s$ in an orthogonal direct sum. Then

$$H(\rho) \leq H\left(\sum_s \mathrm{Proj}_s\,\rho\,\mathrm{Proj}_s\right).$$

Let ρ_j, $1 \leq j \leq m$, be density operators on $\mathcal{H}$, let $\bar{p}$ be a probability distribution on $[m]$, and let $\sum_j p_j\rho_j \in \mathcal{D}(\mathcal{H})$. Then

$$\sum_j p_j H(\rho_j) \leq H\left(\sum_j p_j\rho_j\right) \leq \sum_j p_j H(\rho_j) + H(\bar{p}) \tag{6.2.2}$$

where equality holds in the second inequality if and only if the supports of the ρ_j are all orthogonal.

Let $\rho_{AB} \in \mathcal{D}(\mathcal{H}_A \otimes \mathcal{H}_B)$; then

$$|H(\rho_A) - H(\rho_B)| \leq H(\rho_{AB}) \leq H(\rho_A) + H(\rho_B). \tag{6.2.3}$$

Remark 6.2.21. Von Neuman entropy is nondecreasing under the action of many quantum channels of interest. However, just like with Shannon entropy, it may decrease under certain channels (see Remark 4.3.7), e.g., taking a partial trace. Consider the quantum channel $\Lambda : \mathrm{End}(\mathcal{H}) \to \mathrm{End}(\mathbb{C}^2)$ given by $\rho \mapsto \mathrm{trace}(\rho)|0\rangle\langle 0|$. Then for any $\rho \in \mathcal{D}(\mathcal{H})$ with $H(\rho) > 0$, one obtains $H(\Lambda(\rho)) = 0 < H(\rho)$.

6.3. Conditional von Neumann entropy and strong subadditivity

Recall the conditional Shannon entropy (Definition 4.5.1): the entropy of $p_{\mathcal{X}}$ conditioned on $y = j$, averaged over $j \in \mathcal{Y}$. It is not clear how to "condition" one density matrix on another. Nevertheless the equation $H(\bar{p}_{\mathcal{X}}|\bar{p}_{\mathcal{Y}}) = H(\bar{p}_{\mathcal{X}\times\mathcal{Y}}) - H(\bar{p}_{\mathcal{Y}})$ makes sense for density operators, so we use it as a definition in §6.3.1. This turns out to have unexpected consequences; in particular, the conditional von Neumann entropy may be negative, unlike its classical cousin. Nevertheless, vestiges of positivity remain, as illustrated by the strong subadditivity theorem presented in §6.3.2. Strong subadditivity is proved in §6.3.3 and §6.3.4. The key to the proof is a convexity result.

6.3.1. Conditional von Neumann entropy.

Definition 6.3.1. For a density operator $\rho_{AB} \in \mathcal{D}(\mathcal{H}_A \otimes \mathcal{H}_B)$, the *conditional von Neumann entropy* is

$$H(\rho_A|\rho_B) := H(\rho_{AB}) - H(\rho_B). \tag{6.3.1}$$

WARNING. It is possible that the conditional von Neumann entropy is *negative*. Consider the following example: let $|\mathrm{epr}\rangle = \frac{1}{\sqrt{2}}(|00\rangle + |11\rangle) \in \mathcal{H}_A \otimes \mathcal{H}_B$ and let $\rho_{AB} = |\mathrm{epr}\rangle\langle\mathrm{epr}|$. Then

$$\rho_B = \frac{1}{2}\,\mathrm{Id}_{\mathcal{H}_B} = \frac{1}{2}(|0\rangle\langle 0| + |1\rangle\langle 1|),$$

so $H(\rho_B) = 1$, but $H(\rho_{AB}) = 0$ because ρ_{AB} is pure.

6.3.2. Strong subadditivity.

Theorem 6.3.2 (Strong subadditivity [**LR73**]). *Let $\rho_{ABC} \in \mathcal{D}(\mathcal{H}_A \otimes \mathcal{H}_B \otimes \mathcal{H}_C)$ with marginals ρ_A, ρ_{AB}, etc. Then*

$$H(\rho_C|\rho_A) + H(\rho_C|\rho_B) \geq 0. \tag{6.3.2}$$

In other words,

$$H(\rho_{AC}) - H(\rho_A) + H(\rho_{BC}) - H(\rho_B) \geq 0. \tag{6.3.3}$$

.

Theorem 6.3.2 is proved in §6.3.3.

We may rephrase (6.3.2) in several ways. Here are exercises illustrating some of them:

Exercise 6.3.3. Let ρ_{ABC} be a density operator on $\mathcal{H}_A \otimes \mathcal{H}_B \otimes \mathcal{H}_C$. Then

$$H(\rho_{ABC}) + H(\rho_B) \leq H(\rho_{AB}) + H(\rho_{BC}). \tag{6.3.4}$$

⊚

Exercise 6.3.4 (Subadditivity of conditional entropy).

(1) Let ρ_{ABCD} be a density operator on $\mathcal{H}_A \otimes \mathcal{H}_B \otimes \mathcal{H}_C \otimes \mathcal{H}_D$. Then

$$H(\rho_{AB}|\rho_{CD}) \leq H(\rho_A|\rho_C) + H(\rho_B|\rho_D).$$

(2) Let ρ_{ABC} be a density operator on $\mathcal{H}_A \otimes \mathcal{H}_B \otimes \mathcal{H}_C$. Then

$$H(\rho_{AB}|\rho_C) \leq H(\rho_A|\rho_C) + H(\rho_B|\rho_C) \text{ and}$$
$$H(\rho_A|\rho_{BC}) \leq H(\rho_A|\rho_B) + H(\rho_A|\rho_C).$$

⊚

6.3.3. Proof of strong subadditivity. The proof is based on concavity results. The key result is Corollary 6.3.11 proved below, that the conditional entropy $H(\rho_A|\rho_B)$ is concave in ρ_{AB}. Using that, the proof follows easily:

Proof of Theorem 6.3.2. Let $f(\rho_{ABC}) = H(\rho_C|\rho_A) + H(\rho_C|\rho_B)$, which is a concave function of ρ_{ABC} because it is a sum of two concave functions of ρ_{ABC}. Write

$$\rho_{ABC} = \sum \lambda_i |v_i\rangle\langle v_i|,$$

for some orthogonal unit vectors $|v_i\rangle \in \mathcal{H}_A \otimes \mathcal{H}_B \otimes \mathcal{H}_C$. The concavity of f implies $f(\rho_{ABC}) \geq \sum \lambda_i f(|v_i\rangle\langle v_i|)$. But $\sum \lambda_i f(|v_i\rangle\langle v_i|) = 0$ because, since $|v_i\rangle\langle v_i|$ is pure, $H((|v_i\rangle\langle v_i|)_{AC}) = H((|v_i\rangle\langle v_i|)_B)$ and $H((|v_i\rangle\langle v_i|)_{BC}) = H((|v_i\rangle\langle v_i|)_A)$. We conclude $f(\rho_{ABC}) \geq 0$. □

Remark 6.3.5. One may prove strong subadditivity using representation theory instead of concavity; see §7.4.3.

6.3.4. Lieb's lemma and consequences. Here is the main concavity lemma from which all the others will follow:

Definition 6.3.6. Let $D \subset \mathbb{R}^N$ be a convex subset. A function $f : D \times D \to \mathbb{R}$ is *jointly concave* in its arguments if for all $\lambda_1, \lambda_2 \geq 0$ with $\lambda_1 + \lambda_2 = 1$,

$$f(\lambda_1 x_1 + \lambda_2 x_2, \lambda_1 y_1 + \lambda_2 y_2) \geq \lambda_1 f(x_1, y_1) + \lambda_2 f(x_2, y_2).$$

For a positive operator $A \in \text{End}(\mathcal{H})$ and $p \in \mathbb{C}$, define an operator A^p as follows: by positivity we may write $A = PDP^{-1}$ with D diagonal with nonnegative entries $\lambda_1, \dots, \lambda_n$, and $A^p := PD^pP^{-1}$.

Lemma 6.3.7 (Lieb [**Lie73**]). *Let $X \in \text{End}(\mathcal{H})$. For all $0 \leq p \leq 1$,*

$$f(A, B) := \text{trace}(A^p X^\dagger B^{1-p} X),$$

defined on pairs of positive operators, is jointly concave.

Proof. We follow the proof in [**Rus04**]. First note that it is sufficient to prove the case $A = B$ because

$$\text{trace}(A^p X^\dagger B^{1-p} X) = \text{trace}\begin{pmatrix} A & 0 \\ 0 & B \end{pmatrix}^p \begin{pmatrix} 0 & X^\dagger \\ 0 & 0 \end{pmatrix} \begin{pmatrix} A & 0 \\ 0 & B \end{pmatrix}^{1-p} \begin{pmatrix} 0 & 0 \\ X & 0 \end{pmatrix}.$$

We need to show, for $\lambda_j \geq 0$ and $\lambda_1 + \lambda_2 = 1$, that

$$\begin{aligned}(6.3.5)\qquad &\lambda_1 \operatorname{trace}(A_1^p X^\dagger A_1^{1-p} X) + \lambda_2 \operatorname{trace}(A_2^p X^\dagger A_2^{1-p} X)\\ &\leq \operatorname{trace}\left[(\lambda_1 A_1 + \lambda_2 A_2)^p X^\dagger (\lambda_1 A_1 + \lambda_2 A_2)^{1-p} X\right].\end{aligned}$$

Note that the inequality (in fact equality) holds when $p = 0, 1$. The result would follow if we had a way to show that the right-hand side minus the left-hand side, as a function of p on $[0, 1]$, has its minimum on the boundary. We will make use of a trick to obtain a result of this kind. First we introduce notation to facilitate the calculation:

Assume $C := \lambda_1 A_1 + \lambda_2 A_2$ is invertible; the general case will follow by continuity. Write

$$M = C^{\frac{1-p}{2}} X C^{\frac{p}{2}},$$

which makes sense as C is positive. For $k \in \{1, 2\}$, set

$$(6.3.6)\qquad f_k(p) = \operatorname{trace}(A_k^p C^{-\frac{p}{2}} M^\dagger C^{-\frac{1-p}{2}} A_k^{1-p} C^{-\frac{1-p}{2}} M C^{-\frac{p}{2}});$$

then (6.3.5) becomes

$$(6.3.7)\qquad \lambda_1 f_1(p) + \lambda_2 f_2(p) \leq \operatorname{trace}(M^\dagger M).$$

We need to prove that (6.3.7) holds for $0 \leq p \leq 1$. Here is the promised trick, which is a standard application of elementary complex analysis: extend $f_j(p)$ to be defined for complex numbers on the strip $0 \leq \operatorname{Re}(z) \leq 1$. Why would we want to do this? The maximum principle (see, e.g., [**Ahl78**, §4.3.4]) implies that if $f(z)$ is uniformly bounded on a region, then the maximum of $f(z)$ is achieved on the boundary of the region. So we will trade proving that (6.3.7) holds on the interval $(0, 1)$ to proving that $\lambda_1 f_1(p) + \lambda_2 f_2(p)$ is uniformly bounded on a strip and proving the inequality (6.3.7) on the lines $0 + iy$ and $1 + iy$ where we will have more control over the operators.

It will be convenient to use the operator norm (cf. Definition 3.5.3). Recall that $||X||_{\mathrm{op}}$ is the largest absolute value of an eigenvalue of $X \in \operatorname{End}(\mathcal{H})$.

To see why we might have such control, if A is Hermitian, so $A = \sum \lambda_j |\psi_j\rangle\langle\psi_j|$ with $\lambda_j \in \mathbb{R}$ and $|\psi_j\rangle$ orthonormal, we may define

$$A^w = \sum_j \lambda_j^w |\psi_j\rangle\langle\psi_j|.$$

In particular, if $w = iy$ is imaginary, then $A^{iy} = \sum_j e^{iy\ln(\lambda_j)} |\psi_j\rangle\langle\psi_j|$ is a unitary operator on $\operatorname{supp}(A)$, and the eigenvalues of A^{iy} have norm one. More generally, for $z = x + iy$, $A^z = A^x A^{iy}$ with $||A^{iy}||_{\mathrm{op}} = 1$.

We will utilize that for $X, Y \in \operatorname{End}(\mathcal{H})$ with Y positive,

$$(6.3.8)\qquad |\operatorname{trace}(XY)| \leq ||X||_{\mathrm{op}} \operatorname{trace}(Y).$$

Write λ_{A_k} for the largest eigenvalue of A_k. The inequality (6.3.8) applied to (6.3.6) gives

$$\begin{aligned}|f_k(z)| &\leq \lambda_{A_k}^{z} \operatorname{trace}(C^{-\frac{z}{2}} M^\dagger C^{-\frac{1-z}{2}} A_k^{1-z} C^{-\frac{1-z}{2}} M C^{-\frac{z}{2}}) \\ &\leq \lambda_{A_k}^{z} \lambda_{A_k}^{1-z} \operatorname{trace}(C^{-\frac{z}{2}} M^\dagger C^{-\frac{1-z}{2}} C^{-\frac{1-z}{2}} M C^{-\frac{z}{2}}) \\ &\leq \|A_k\|_{\mathrm{op}} \|C^{-1}\|_{\mathrm{op}} \operatorname{trace}(M^\dagger M),\end{aligned}$$

so $f_k(z)$ is indeed uniformly bounded on the strip. It remains to show that (6.3.7) holds on the boundary. Consider

$$\begin{aligned} f_k(0+iy) &= \operatorname{trace}(A_k^{iy} C^{-\frac{iy}{2}} M^\dagger C^{\frac{iy}{2}} C^{-\frac{1}{2}} A_k^{1-iy} C^{-\frac{iy}{2}} C^{-\frac{1}{2}} M C^{-\frac{iy}{2}}) \\ &= \operatorname{trace}\left((A_k^{\frac{iy}{2}} C^{-\frac{iy}{2}} M^\dagger C^{\frac{iy}{2}} C^{-\frac{1}{2}} A_k^{\frac{1}{2}})(A_k^{\frac{1}{2}} A_k^{-iy} A_k^{\frac{1}{2}} C^{-\frac{iy}{2}} C^{-\frac{1}{2}} M C^{-\frac{iy}{2}} A_k^{\frac{iy}{2}})\right) \end{aligned} \tag{6.3.9}$$

where we used $A_k^{iy} = A_k^{\frac{iy}{2}} A_k^{\frac{iy}{2}}$, $\operatorname{trace}(ZW) = \operatorname{trace}(WZ)$, and split $A_k^{1-iy} = A_k^{\frac{1}{2}} A_k^{\frac{1}{2}} A_k^{-iy}$.

The Cauchy-Schwartz inequality $|\langle u|v\rangle|^2 \leq |u|^2|v|^2$ implies $|\operatorname{trace}(ZW)| \leq \left[\operatorname{trace}(ZZ^\dagger)\operatorname{trace}(WW^\dagger)\right]^{\frac{1}{2}}$. Take Z to be the first term in parentheses in (6.3.9) and take W to be the second. We have

$$\begin{aligned}&|f_k(0+iy)| \\ &\leq \left[\operatorname{trace}\left((A_k^{\frac{iy}{2}} C^{-\frac{iy}{2}} M^\dagger C^{\frac{iy}{2}} C^{-\frac{1}{2}} A_k^{\frac{1}{2}})(A_k^{\frac{1}{2}} C^{\frac{1}{2}} C^{\frac{iy}{2}} M C^{\frac{iy}{2}} A_k^{\frac{-iy}{2}})\right)\right]^{\frac{1}{2}} \\ &\quad * \left[\operatorname{trace}\left((A_k^{\frac{1}{2}} A_k^{-iy} A_k^{\frac{1}{2}} C^{-\frac{iy}{2}} C^{-\frac{1}{2}} M C^{\frac{iy}{2}} A_k^{-\frac{iy}{2}})(A_k^{\frac{iy}{2}} C^{\frac{-iy}{2}} M^\dagger C^{-\frac{1}{2}} C^{\frac{iy}{2}} A_k^{\frac{1}{2}} A_k^{iy} A_k^{\frac{1}{2}})\right)\right]^{\frac{1}{2}} \\ &= \left[\operatorname{trace}(M^\dagger C^{\frac{iy}{2}} C^{-\frac{1}{2}} A_k C^{-\frac{1}{2}} C^{-\frac{iy}{2}} M) \operatorname{trace}(A_k^{\frac{1}{2}} C^{-\frac{1}{2}} C^{-\frac{iy}{2}} M M^\dagger C^{-\frac{iy}{2}} C^{-\frac{1}{2}} A_k^{\frac{1}{2}})\right]^{\frac{1}{2}}.\end{aligned}$$

Noting that these two traces are the same, we obtain

$$|f_k(0+iy)| \leq \operatorname{trace}(M^\dagger C^{\frac{iy}{2}} C^{-\frac{1}{2}} A_k C^{-\frac{1}{2}} C^{-\frac{iy}{2}} M).$$

Putting it all together,

$$\begin{aligned}|\lambda_1 f_1(0+iy) + \lambda_2 f_2(0+iy)| &\leq \lambda_1 |f_1(0+iy)| + \lambda_2 |f_2(0+iy)| \\ &\leq \operatorname{trace}(M^\dagger C^{\frac{iy}{2}} C^{-\frac{1}{2}} (\lambda_1 A_1 + \lambda_2 A_2) C^{-\frac{1}{2}} C^{-\frac{iy}{2}} M) \\ &= \operatorname{trace}(M^\dagger M),\end{aligned}$$

proving that (6.3.7) holds on the line $z = 0 + iy$. The case of $f(1+iy)$ is similar. □

Theorem 6.3.8. *The relative entropy $H(\rho||\sigma)$ is jointly convex in ρ, σ.*

Proof. Let $A, X \in \mathrm{End}(\mathcal{H})$ with A positive. Define

$$I_t(A,X) := \mathrm{trace}(X^\dagger A^t X A^{1-t}) - \mathrm{trace}(X^\dagger X A),$$

where the second term is included just to normalize $I_0(X,A) = 0$. Lieb's lemma (Lemma 6.3.7) implies the first term is concave in A, and since the second is linear in A, $I_t(A,X)$ is concave in A. Write $I_0'(A,X) = \frac{d}{dt}|_{t=0} I_t(A,X)$. Recall that $\frac{d}{dt}|_{t=0} a^t = \ln(a)$, and the same holds for positive operators, so

$$I_0'(A,X) = \mathrm{trace}(X^\dagger \ln(A) X A) - \mathrm{trace}(X^\dagger X \ln(A) A).$$

In particular,

$$\begin{aligned} I_0'\left(\begin{pmatrix}\rho & 0\\ 0 & \sigma\end{pmatrix}, \begin{pmatrix}0 & 0\\ \mathrm{Id} & 0\end{pmatrix}\right) &= \mathrm{trace}\left(\begin{pmatrix}0 & \mathrm{Id}\\ 0 & 0\end{pmatrix}\begin{pmatrix}\ln(\rho) & 0\\ 0 & \ln(\sigma)\end{pmatrix}\begin{pmatrix}0 & 0\\ \mathrm{Id} & 0\end{pmatrix}\begin{pmatrix}\rho & 0\\ 0 & \sigma\end{pmatrix}\right)\\ &\quad - \mathrm{trace}\left(\begin{pmatrix}0 & \mathrm{Id}\\ 0 & 0\end{pmatrix}\begin{pmatrix}0 & 0\\ \mathrm{Id} & 0\end{pmatrix}\begin{pmatrix}\ln(\rho) & 0\\ 0 & \ln(\sigma)\end{pmatrix}\begin{pmatrix}\rho & 0\\ 0 & \sigma\end{pmatrix}\right)\\ &= -\frac{H(\rho||\sigma)}{\ln(2)}, \end{aligned}$$

so it suffices to prove that $I_0'(A,X)$ is a concave function of A. Now

$$\begin{aligned} I_0'(\lambda_1 A_1 + \lambda_2 A_2, X) &= \lim_{\epsilon\to 0} \frac{I_\epsilon(\lambda_1 A_1 + \lambda_2 A_2, X) - 0}{\epsilon}\\ &\geq \lim_{\epsilon\to 0} \lambda_1 \frac{I_\epsilon(A_1,X)}{\epsilon} + \lambda_2 \frac{I_\epsilon(A_2,X)}{\epsilon} \qquad \text{by concavity}\\ &= \lambda_1 I_0'(A_1,X) + \lambda_2 I_0'(A_2,X), \end{aligned}$$

so $I_0'(A,X)$ is indeed a concave function of A and we conclude $H(\rho||\sigma)$ is jointly convex in ρ, σ. □

Exercise 6.3.9. Let $X \in \mathrm{End}(\mathbb{C}^n)$ be Hermitian and of rank n. Show that there exists a mixed unitary channel (Example 5.5.25) $\Lambda : \mathrm{End}(\mathbb{C}^n) \to \mathrm{End}(\mathbb{C}^n)$ such that $\Lambda(X) = \frac{1}{n}\mathrm{trace}(X)\,\mathrm{Id}_{\mathbb{C}^n}$; i.e., there exists a probability distribution p_j and $U_j \in \mathbf{U}_n$ such that

$$\sum_i p_i U_i X U_i^{-1} = \frac{1}{n}\mathrm{trace}(X)\,\mathrm{Id}_{\mathbb{C}^n}. \qquad ⊚$$

Exercise 6.3.10 (Monotonicity of the relative entropy)**.** Let ρ_{AB}, σ_{AB} be density operators on $\mathcal{H}_A \otimes \mathcal{H}_B$. Show that

$$H(\rho_A||\sigma_A) \leq H(\rho_{AB}||\sigma_{AB}). \qquad ⊚$$

Exercise 6.3.10 may be interpreted as saying that if one ignores part of a physical system, it does not make it easier to distinguish the state of the system.

Corollary 6.3.11. *The conditional entropy $H(\rho_A|\rho_B)$ is concave in ρ_{AB}.*

Proof. Let $d = \dim \mathcal{H}_A$. Consider

$$H\left(\rho_{AB} \middle\| \frac{1}{d} \mathrm{Id}_{\mathcal{H}_A} \otimes \rho_B\right) = -H(\rho_{AB}) - \mathrm{trace}_{\mathcal{H}_A \otimes \mathcal{H}_B}\left(\rho_{AB} \log\left(\frac{1}{d} \mathrm{Id}_{\mathcal{H}_A} \otimes \rho_B\right)\right).$$

Exercise 6.2.12 implies that

$$\begin{aligned} \mathrm{trace}_{\mathcal{H}_A \otimes \mathcal{H}_3}\left(\rho_{AB} \log\left(\frac{1}{d} \mathrm{Id}_{\mathcal{H}_A} \otimes \rho_B\right)\right) &= \mathrm{trace}(\rho_B \log(\rho_B)) + \log(d) \\ &= -H(\rho_B) + \log(d). \end{aligned}$$

We have

$$\begin{aligned} H(\rho_A|\rho_B) &= H(\rho_{AB}) - H(\rho_B) \\ &= -H\left(\rho_{AB} \middle\| \frac{1}{d} \mathrm{Id}_{\mathcal{H}_A} \otimes \rho_B\right) - [-H(\rho_B) + \log(d)] - H(\rho_B) \\ &= -\log(d) - H\left(\rho_{AB} \middle\| \frac{1}{d} \mathrm{Id}_{\mathcal{H}_A} \otimes \rho_B\right). \end{aligned}$$

Finally, the last line is jointly concave in ρ_{AB} and $\frac{1}{d} \mathrm{Id}_{\mathcal{H}_A} \otimes \rho_B$ by Theorem 6.3.8. □

6.4. Information-theoretic consequences of strong subadditivity

6.4.1. Quantum mutual information.

As with the classical case, we have three expressions for the mutual information, each of which has a quantum-informational interpretation:

Definition 6.4.1. Define the *quantum mutual information* of

$$\rho_{AB} \in \mathcal{D}(\mathcal{H}_A \otimes \mathcal{H}_B)$$

with marginals ρ_A, ρ_B to be

$$\begin{aligned} I(\rho_A : \rho_B) :&= H(\rho_A) + H(\rho_B) - H(\rho_{AB}) \\ &= H(\rho_A) - H(\rho_A|\rho_B) \\ &= H(\rho_B) - H(\rho_B|\rho_A). \end{aligned}$$

Here is another consequence of strong subadditivity:

Corollary 6.4.2 (Mutual information is nonincreasing when quantum systems are discarded). *Let ρ_{ABC} be a density operator on $\mathcal{H}_A \otimes \mathcal{H}_B \otimes \mathcal{H}_C$. Then*

$$I(\rho_A : \rho_B) \leq I(\rho_A : \rho_{BC}).$$

Exercise 6.4.3. Write out the proof of Corollary 6.4.2.

The following is a quantum analog of Exercise 4.5.7, the classical data processing inequality:

Proposition 6.4.4 (Quantum data processing inequality). *Assume that* $\rho_{AB} \in \mathcal{D}(\mathcal{H}_A \otimes \mathcal{H}_B)$, *and let* $\Lambda : \mathrm{End}(\mathcal{H}_B) \to \mathrm{End}(\mathcal{H}_C)$ *be a quantum channel. Let* $\tilde{\Lambda} = \mathrm{Id}_{\mathcal{H}_A} \otimes \Lambda : \mathrm{End}(\mathcal{H}_A \otimes \mathcal{H}_B) \to \mathrm{End}(\mathcal{H}_A \otimes \mathcal{H}_C)$. *Then*

$$I(\rho_A : \rho_B) \geq I(\tilde{\Lambda}(\rho_{AB})_A : \tilde{\Lambda}(\rho_{AB})_C).$$

Proposition 6.4.4 may be interpreted as saying that one cannot increase mutual information ("correlations") by acting locally.

Proof. Take a Stinespring representation $\Lambda(X) = \mathrm{trace}_{\mathcal{H}_E}(ZXZ^\dagger)$ for some isometry $Z : \mathcal{H}_B \to \mathcal{H}_C \otimes \mathcal{H}_E$ and define

$$\sigma_{ACE} := (\mathrm{Id}_{\mathcal{H}_A} \otimes Z)\rho_{AB}(\mathrm{Id}_{\mathcal{H}_A} \otimes Z)^\dagger.$$

Observe that $\mathrm{trace}_{\mathcal{H}_E}(\sigma_{ACE}) = \tilde{\Lambda}(\rho_{AB})$. Thus

$$\begin{aligned} I(\rho_A : \rho_B) &= I(\sigma_A : \sigma_{CE}) \\ &\geq I(\tilde{\Lambda}(\rho_{AB})_A : \tilde{\Lambda}(\rho_{AB})_C), \end{aligned}$$

the first line by invariance of mutual information under isometry and the second by Corollary 6.4.2. □

6.4.2. Classical communication along a quantum channel. A variant of the data processing inequality is *Holevo's bound*, which upper bounds the amount of classical information that can be extracted from a quantum state, or the capacity of a quantum channel to transmit classical information. This will include both the cases of a noiseless channel (where Λ is an isometry) and a channel with noise.

Recall the set-up: Alice wants to transmit classical messages to Bob using a quantum channel. That is, a classical source $\mathcal{X}$ emits symbols x, according to a probability distribution $p_{\mathcal{X}}$, which Alice encodes as density operators ρ_x, the operators ρ_x are sent to Bob along a quantum channel, and then Bob (who knows Alice's dictionary) performs a measurement to obtain a classical message (a basis vector).

To facilitate the mixing of classical and quantum, let $\mathcal{H}_{\mathcal{X}} := \mathbb{C}^{|\mathcal{X}|}$ with basis elements $|j\rangle$ indexed by $j \in \mathcal{X}$. Define

$$\rho_{\mathcal{X}B} := \sum_{j \in \mathcal{X}} p_j |j\rangle\langle j| \otimes \rho_j \in \mathcal{D}(\mathcal{H}_{\mathcal{X}} \otimes \mathcal{H}_B).$$

The mutual information $I(\rho_{\mathcal{X}} : \rho_B)$ turns out to be an important quantity so it gets its own name:

Definition 6.4.5. The *Holevo* χ*-quantity* is

$$\chi(\rho_{\mathcal{X}B}) := I(\rho_{\mathcal{X}} : \rho_B).$$

We have

$$\begin{aligned}\chi(\rho_{\mathcal{X}B}) &= H(\rho_{\mathcal{X}}) + H(\rho_B) - H(\rho_{\mathcal{X}B}) \\ &= H(\bar{p}_{\mathcal{X}}) + H\left(\sum_j p_j\rho_j\right) - \left[H(\bar{p}_{\mathcal{X}}) + \sum_j p_jH(\rho_j)\right] \\ &= H\left(\sum_j p_j\rho_j\right) - \sum_j p_jH(\rho_j).\end{aligned}$$

The second line is a consequence of Exercise 6.2.17.

In particular, $0 \le \chi(\rho_{\mathcal{X}B}) \le H(\sum_j p_j\rho_j)$, where the first inequality is due to concavity of the von Neumann entropy (Proposition 6.2.18).

Exercise 6.4.6. Show that $\chi(\rho_{\mathcal{X}B}) \le H(\bar{p}_{\mathcal{X}})$ with equality if and only if the states ρ_j with $p_j > 0$ have pairwise orthogonal support.

Theorem 6.4.7 (Holevo bound). *No matter what measurement channel*

$$\Lambda : \mathrm{End}(\mathcal{H}_B) \to \mathrm{End}(\mathcal{H}_y)$$

Bob applies to ρ_B,

$$I(\rho_{\mathcal{X}} : \rho_y) \le \chi(\rho_{\mathcal{X}B}).$$

Proof. Set $\tilde{\Lambda} = \mathrm{Id}_{\mathcal{H}_{\mathcal{X}}} \otimes \Lambda$. The data processing inequality (Proposition 6.4.4) implies that

$$\chi(\rho_{\mathcal{X}B}) = I(\rho_{\mathcal{X}} : \rho_B) \ge I(\tilde{\Lambda}(\rho)_{\mathcal{X}} : \tilde{\Lambda}(\rho)_y) = I(\rho_{\mathcal{X}} : \rho_y). \qquad \square$$

Exercise 6.4.8. Use Exercise 6.4.6 to show that, assuming $p_{\mathcal{X}}(j) > 0$ for all $j \in \mathcal{X}$, information can only be exactly transmitted if $\dim \mathcal{H}_B \ge |\mathcal{X}|$.

Exercise 6.4.8 says that exact transmission of m bits requires the transmission of at least m qubits.

If instead of exact transmission, one is concerned with reliable transmission, the capacity is given by Theorem 6.4.10:

Definition 6.4.9. The *Holevo capacity of a quantum channel* $\Lambda_{AB} : \mathrm{End}(\mathcal{H}_A) \to \mathrm{End}(\mathcal{H}_B)$ is

$$\chi(\Lambda_{AB}) := \sup_{\rho_{A\mathcal{X}} \in \mathcal{D}(\mathcal{H}_A \otimes \mathcal{H}_B)} \chi((\Lambda_{AB} \otimes \mathrm{Id}_{\mathcal{X}})(\rho_{\mathcal{X}A})).$$

Theorem 6.4.10 (Holevo-Schumacher-Westmorland theorem). *The classical capacity of a quantum channel $\Lambda : \mathrm{End}(\mathcal{H}_A) \to \mathrm{End}(\mathcal{H}_B)$, i.e., the average number of classical bits that can be reliably transmitted with a single use of Λ, is*

$$\lim_{k\to\infty} \frac{\chi(\Lambda^{\otimes k})}{k}.$$

See, e.g., [**Wat18**, Chap. 8.2] for a proof.

6.4.3. Quantum communication along a quantum channel. In order to define the quantum capacity of a quantum channel, first define the *coherent information of* $\rho \in \mathcal{D}(\mathcal{H}_A)$ *through a channel* $\Lambda : \mathrm{End}(\mathcal{H}_A) \to \mathrm{End}(\mathcal{H}_B)$ to be

$$I_{\mathrm{Coh}}(\rho;\Lambda) := H(\Lambda(\rho)) - H((\Lambda \otimes \mathrm{Id}_{\mathcal{H}_A})(|\sqrt{\rho}\rangle\langle\sqrt{\rho}|)),$$

where $|\sqrt{\rho}\rangle\langle\sqrt{\rho}|$ is the standard purification of ρ. Next optimize over choices of density operator to define the *coherent information of* Λ:

$$I_{\mathrm{Coh}}(\Lambda) := \max_{\rho \in \mathcal{D}(\mathcal{H}_A)} I_{\mathrm{Coh}}(\rho;\Lambda).$$

Theorem 6.4.11. *Assume notations as above. The quantum capacity of a quantum channel* Λ*, i.e., the average number of qubits that can be reliably transmitted with a single use of* Λ*, is*

$$\lim_{k\to\infty} \frac{I_{\mathrm{Coh}}(\Lambda^{\otimes k})}{k}.$$

See [**Wat18**, §8.2] for the proof.

Remark 6.4.12. For readers familiar with the asymptotic rank and subrank of tensors, the capacities in Theorems 6.4.10 and 6.4.11 may be compared with the logs of the asymptotic ranks and subranks:

$$\lim_{k\to\infty} \frac{1}{k}\log(Q(T^{\boxtimes k})) \quad \text{and} \quad \lim_{k\to\infty} \frac{1}{k}\log(R(T^{\boxtimes k})).$$

6.5. Entanglement and LOCC

We have seen several ways that *entanglement* is already a resource for the space $\mathcal{H}_A \otimes \mathcal{H}_B = \mathbb{C}^2 \otimes \mathbb{C}^2$: using a pre-shared qubit $|\mathrm{epr}\rangle = \frac{1}{\sqrt{2}}(|00\rangle + |11\rangle)$ one can:

(i) Transport two bits of classical information by: Alice acts unitarily on her part of the system, transmits her state to Bob, and Bob acts unitarily on his part and then measures ("super-dense coding", §2.4).

(ii) Transmit one qubit of quantum information from Alice to Bob by: Alice acts unitarily on her two qubits (the first of which she wants Bob to gain possession of), performs a measurement, sends two classical bits to Bob, and then Bob acts unitarily on his qubit to obtain possession of Alice's qubit ("teleportation", §2.5).

In this section we study entanglement as a resource. Unlike quantities such as quantum channel capacity and von Neumann entropy, there is no obvious classical cousin of entanglement. For pure states $|\psi\rangle\langle\psi| \in \mathrm{End}(\mathcal{H}_1 \otimes \cdots \otimes \mathcal{H}_n)$, in Definition 2.3.18, ψ was defined to be entangled if it is not a rank one tensor, i.e., not of the form $|\psi_1\rangle \otimes \cdots \otimes |\psi_n\rangle$. We have not yet defined what it means for a mixed state to be entangled. This will be rectified by Definition 6.6.5.

Assume several different laboratories can communicate classically, have prepared a shared state in advance, and can perform unitary and projection operations on their parts of the states. This was the situation for teleportation. More precisely, make the following assumptions:

- $\mathcal{H} = \mathcal{H}_1 \otimes \cdots \otimes \mathcal{H}_k$, and the $\mathcal{H}_j$ share an entangled state ρ. Primarily we will just have $\mathcal{H} = \mathcal{H}_A \otimes \mathcal{H}_B$.
- The laboratories can communicate classically.
- Each laboratory is allowed to perform unitary and measurement operations on its own space.

The above assumptions are called *LOCC* for "local operations and classical communication".

Restrict to the case $\mathcal{H} = \mathcal{H}_A \otimes \mathcal{H}_B$, with $\mathcal{H}_A, \mathcal{H}_B$ of dimension n. In this section we will be concerned with measuring the "value" of a given entangled state. We will use $|\mathrm{epr}\rangle$ as a benchmark for measuring the quality of entanglement.

We will not be concerned with a single state ρ, but with the tensor product of many copies of it, $\rho^{\otimes d} \in \mathcal{D}((\mathcal{H}_A \otimes \mathcal{H}_B)^{\otimes d})$. We ask "how much" entanglement does $\rho^{\otimes d}$ have? More precisely, what is the function $\delta(d)$ such that we may construct $(|\mathrm{epr}\rangle\langle\mathrm{epr}|)^{\otimes \delta(d)}$ from $\rho^{\otimes d}$ via LOCC? We will be content with output that is close to $(|\mathrm{epr}\rangle\langle\mathrm{epr}|)^{\otimes \delta(d)}$, i.e., has high fidelity (Definition 5.3.4) with $(|\mathrm{epr}\rangle\langle\mathrm{epr}|)^{\otimes \delta(d)}$.

The unitary and projection local operators Alice is allowed to apply amount to a mixed unitary channel (Example 5.5.25):

$$\rho_A \mapsto \sum_{s=1}^{k} p_s U_s \rho_A U_s^{-1} \tag{6.5.1}$$

where the U_s are unitary and p is a probability distribution on $\{1, \ldots, k\}$ for some finite k, and the same holds for Bob. When Alice performs a measurement she also gets the result, which she may send classically to Bob.

The results we prove will only be for pure states.

6.5.1. A classical cousin of the mixed unitary channel. To gain insight as to which states can be produced via LOCC from a given density operator, we examine a classical cousin of (6.5.1) by considering diagonal density operators. This will consist of averaging our probability measure over some probability distribution on elements of the permutation group $\mathfrak{S}_n$:

$$\bar{p} \mapsto \sum_{\sigma \in \mathfrak{S}_n} q_\sigma \mu(\sigma) \bar{p}, \tag{6.5.2}$$

where for a permutation $\sigma \in \mathfrak{S}_n$, $\mu(\sigma)$ is the corresponding permutation matrix and q is a probability distribution on $\mathfrak{S}_n$. Let $M_q = \sum_{\sigma \in \mathfrak{S}_n} q_\sigma \mu(\sigma)$ denote the matrix representing this linear map $\mathbb{R}^n \to \mathbb{R}^n$.

Exercise 6.5.1. Show that M_q is doubly stochastic. ⊚

Recall from Exercise 4.3.5 that Shannon entropy is nondecreasing under an action of the form (6.5.2).

What does the set of all possible M_q's look like? It is some subset of the vector space of all $n \times n$-matrices. It contains the $n!$ permutation matrices and all convex linear combinations of them. In other words, the set of the M_q's is the convex hull of the permutation matrices, $\mathrm{conv}(\mu(\mathfrak{S}_n))$. We have observed that $\mathrm{conv}(\mu(\mathfrak{S}_n)) \subseteq \mathcal{DS}_n$, where $\mathcal{DS}_n \subset \mathrm{Mat}_{n\times n}(\mathbb{R})$ denotes set of doubly stochastic matrices. In fact, equality holds:

Theorem 6.5.2 (Birkhoff [**Bir46**]). $\mathcal{DS}_n = \mathrm{conv}(\mu(\mathfrak{S}_n))$.

That is, the set of doubly stochastic matrices is a convex polytope with vertices the permutation matrices. We outline the proof of Theorem 6.5.2 in §6.5.2 below.

Define a partial order on probability distributions on $[n]$, by $\bar{q} \leq \bar{p}$ if

$$\bar{q} \in \mathcal{DS}_n \cdot \bar{p}.$$

A classical theorem of Hardy-Littlewood-Polya states that this partial order coincides with the *dominance order*:

Definition 6.5.3. Let $x, y \in \mathbb{R}^n$ and write $x^{\downarrow}$ for x re-ordered such that $x_1 \geq x_2 \geq \cdots \geq x_n$. We say $x \prec y$ if for all $k \leq n$,

$$\sum_{j=1}^{k} x_j^{\downarrow} \leq \sum_{j=1}^{k} y_j^{\downarrow}.$$

The partial order determined by $\prec$ is called the *dominance order* or *majorization.*

If p is a probability distribution concentrated at a point, then $\bar{q} \prec \bar{p}$ for all probability distributions q, and if p is such that $p_j = \frac{1}{n}$ for all j, then $\bar{p} \prec \bar{q}$ for all q. We will also see that the dominance order is compatible with the entropy in the sense that $\bar{p} \prec \bar{q}$ implies $H(\bar{p}) \geq H(\bar{q})$.

Theorem 6.5.4 (Hardy-Littlewood-Pólya [**HLP52**]). *Fix* $y \in \mathbb{R}^n$. *Then*

$$\{x \in \mathbb{R}^n \mid x \prec y\} = \mathcal{DS}_n \cdot y.$$

We outline the proof of Theorem 6.5.4 in §6.5.2.

6.5.2. Proofs of Theorems 6.5.2 and 6.5.4.

We follow [**Mir58**].

Exercise 6.5.5. Fix $y \in \mathbb{R}^n$. Show that the set $\{x \mid x \prec y\}$ is convex. ⊚

Lemma 6.5.6. *Let* $A \in \mathcal{DS}_n \backslash \mathrm{Id}$. *Then there exists* $\sigma \in \mathfrak{S}_n \backslash \mathrm{Id}$ *such that the diagonal elements of* $\mu(\sigma)A$ *are all nonzero.*

Proof. Say there is no such σ; i.e., for each $\sigma \in \mathfrak{S}_n \backslash \mathrm{Id}$ there exists some j such that $A_{j,\sigma(j)} = 0$. Then

$$\det(A) = \sum_{\sigma \in \mathfrak{S}_n} \operatorname{sgn}(\sigma) A_{1\sigma(1)} \cdots A_{n\sigma(n)} = A_{11} \cdots A_{nn}.$$

Similarly, $\det(A + t\,\mathrm{Id}) = (A_{11} + t) \cdots (A_{nn} + t)$ so the eigenvalues of A are the diagonal elements. But at least one eigenvalue is 1 because of the eigenvector $(1, \ldots, 1)^{\mathbf{t}}$. Say this is A_{jj}. Then strike out the j-th row and column of A and apply the argument again. Continuing, we see $A = \mathrm{Id}$, a contradiction. □

Exercise 6.5.7. Let $x, y_1, \ldots, y_m \in \mathbb{R}^n$. Show that $x \in \operatorname{conv}\{y_1, \ldots, y_m\}$ if and only if for all $\alpha \in \mathbb{R}^{n*}$, $\alpha(x) \leq \max_{s \in \{1,\ldots,m\}} \alpha(y_s)$. ⊚

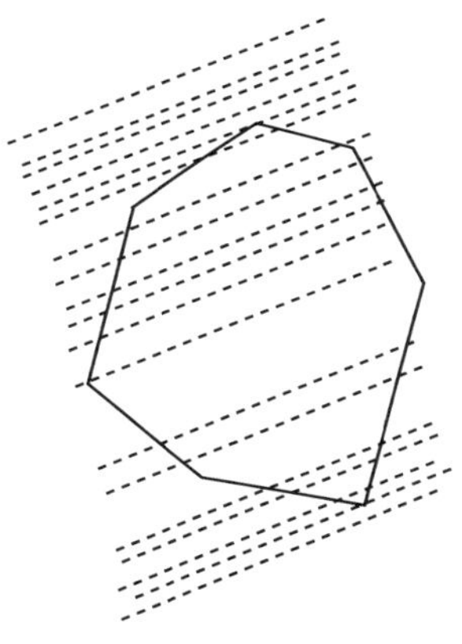

Figure 6.5.1. Level sets of a linear form, max and min of the form on a convex set occur on the boundary

Remark 6.5.8. Exercise 6.5.7 is important for linear optimization problems on a convex set: one just needs to examine the boundary vertices.

Exercise 6.5.9. Use Exercise 6.5.7 to show that $x \prec y$ implies $x \in \operatorname{conv}(\mathcal{DS}_n \cdot y)$.

Exercise 6.5.10. Prove Theorem 6.5.4.

Exercise 6.5.11. Show that for all $A \in \mathrm{Mat}_{n \times n}$,

$$\max_{D \in \mathcal{DS}_n} \operatorname{trace}(DA) = \max_{\sigma \in \mathfrak{S}_n} (\operatorname{trace}(\mu(\sigma) A)). \quad ⊚$$

Exercise 6.5.12. Use Exercise 6.5.7, with $\{y_1, \ldots, y_m\}$ the set of permutation matrices, to prove Theorem 6.5.2.

Exercise 6.5.13. Show that $\bar{p} \prec \bar{q}$ if and only if $\sum_i f(p_i) \leq \sum_i f(q_i)$ for all convex functions $f : [0,1] \to \mathbb{R}$.

Exercise 6.5.14. Use Exercise 6.5.13 to show that $\bar{p} \prec \bar{q}$ implies $H(\bar{p}) \geq H(\bar{q})$.

6.5.3. Dominance order for Hermitian operators. Let X, Y be Hermitian operators and write $\mathrm{spec}(X)$ for the set of eigenvalues of X (the spectrum of X). Define $X \prec Y$ if $\mathrm{spec}(X) \prec \mathrm{spec}(Y)$, where here we slightly abuse notation and consider $\mathrm{spec}(X)$ as a list of the eigenvalues in some order. The order does not matter because of the reordering in Definition 6.5.3.

Exercise 6.3.9 shows that $\frac{1}{n}\,\mathrm{Id} \prec \rho$ for all $\rho \in \mathcal{D}(\mathbb{C}^n)$.

Theorem 6.5.15. *Let X, Y be Hermitian operators on $\mathcal{H} = \mathbb{C}^n$. Then $X \prec Y$ if and only if there exists a mixed unitary channel (Example 5.5.25) Λ such that $\Lambda(Y) = X$; i.e., there exists a probability distribution $\bar{p} = (p_1, \dots, p_{n!})$ and unitary matrices $U_\sigma \in \mathbf{U}(\mathcal{H})$, $1 \le \sigma \le n!$, such that*

$$X = \sum_\sigma p_\sigma U_\sigma Y U_\sigma{}^{-1}. \tag{6.5.3}$$

Proof. Say $X \prec Y$. Theorem 6.5.4 implies that $\mathrm{spec}(X) = \sum_{\sigma \in \mathfrak{S}_n} p_\sigma \mu(\sigma)\,\mathrm{spec}(Y)$ for some probability distribution $\bar{p}$ on $\mathfrak{S}_n$. Write $X = UD(X)U^{-1}$ where U is unitary and $D(X)$ is a diagonal matrix with the eigenvalues of X on the diagonal and similarly $Y = VD(Y)V^{-1}$. By hypothesis $D(X) = \sum_{\sigma \in \mathfrak{S}_n} p_\sigma \mu(\sigma) D(Y) \mu(\sigma)^{-1}$, so

$$X = U^{-1}\left(\sum_\sigma p_\sigma \mu(\sigma) V^{-1} Y V \mu(\sigma)^{-1}\right) U.$$

To conclude, set $U_\sigma = U^{-1}\mu(\sigma)V^{-1}$.

Now say (6.5.3) holds. We have

$$X = UD(X)U^{-1} = \sum_\sigma p_\sigma U_\sigma V D(Y) V^{-1} U_\sigma{}^{-1};$$

i.e.,

$$D(X) = \sum_\sigma p_\sigma (UU_\sigma V) D(Y) (UU_\sigma V)^{-1}.$$

Write $W_\sigma = UU_\sigma V \in \mathbf{U}(\mathcal{H})$. Since $D(X), D(Y)$ are diagonal, conjugation by W_σ must take a diagonal matrix to a diagonal matrix. If the eigenvalues of X are distinct, then W_σ must be a permutation matrix, and in general, without loss of generality, we may assume it to be so. □

Exercise 6.5.16. Show that for all $\rho \in \mathcal{D}(\mathbb{C}^n)$, $\rho \prec |\psi\rangle\langle\psi|$ for each unit length $|\psi\rangle \in \mathbb{C}^n$.

We would like to use the partial order on density operators to make assertions about their value for LOCC. Our actual result will not be as general as this wish. As mentioned above, we will have to restrict to ρ_{AB} being a pure state.

6.5.4. A reduction theorem. The study of LOCC is potentially unmanagable because there can be numerous rounds of local operations and classical communication, making it hard to model. The following result eliminates this problem:

Proposition 6.5.17. *Let $\mathcal{H}_A, \mathcal{H}_B$ be $\mathbb{C}^n$; in particular, they are equipped with bases. If $|\psi\rangle \in \mathcal{H}_A \otimes \mathcal{H}_B$ can be transformed into $|\phi\rangle$ by LOCC, then it can be transformed to $|\phi\rangle$ by the following sequence of operations:*

(1) *Alice performs a single measurement with operators* $\mathrm{Proj}_{\mathcal{M}_s}$.

(2) *She sends the result of her measurement (some s) to Bob classically.*

(3) *Bob performs a unitary operation on his system.*

The sequence of actions in Proposition 6.5.17 may be interpreted as the following type of quantum channel:

$$\begin{aligned}(6.5.4)\quad \Lambda : \mathrm{End}(\mathcal{H}_A \otimes \mathcal{H}_B) &\to \mathrm{End}(\mathcal{H}_A \otimes \mathcal{H}_B)\\ \rho &\mapsto \sum_s (\mathrm{Id}_{\mathcal{H}_A} \otimes U_s)(\mathrm{Proj}_{\mathcal{M}_s} \otimes \mathrm{Id}_{\mathcal{H}_B})\rho(\mathrm{Proj}_{\mathcal{M}_s} \otimes \mathrm{Id}_{\mathcal{H}_B})^\dagger (\mathrm{Id}_{\mathcal{H}_A} \otimes U_s)^\dagger .\end{aligned}$$

Proof. As with the Choi-Jamiołkowski isomorphism (Definition 5.5.9), use the bases to obtain isomorphisms $\mathcal{H}_B^* \to \mathcal{H}_B$, $\langle j_B| \mapsto |j_B\rangle$, and $\mathcal{H}_A^* \to \mathcal{H}_A$. Thus $|\psi\rangle$ induces linear maps $\psi_{BA} : \mathcal{H}_B \to \mathcal{H}_A$ and $\psi_{AB} : \mathcal{H}_A \to \mathcal{H}_B$. The map ψ_{BA} will give the mechanism to transfer Bob's measurements to Alice.

Write $\mathcal{H}_A = \ker \psi_{AB} \oplus \mathcal{H}_A'$ and $\mathcal{H}_B = \ker \psi_{BA} \oplus \mathcal{H}_B'$. Then ψ_{BA} induces an isomorphism $\psi_{BA}' : \mathcal{H}_B' \to \mathcal{H}_A'$, which in turn induces an isomorphism $\tilde{\psi}_{BA}' : \mathrm{End}(\mathcal{H}_B') \to \mathrm{End}(\mathcal{H}_A')$. Extend $\tilde{\psi}_{BA}'$ to a map $\tilde{\psi}_{BA} : \mathrm{End}(\mathcal{H}_B) \to \mathrm{End}(\mathcal{H}_A)$ that acts by zero on the orthogonal complement to $\mathrm{End}(\mathcal{H}_B')$ and then use the inclusion $\mathrm{End}(\mathcal{H}_A') \subset \mathrm{End}(\mathcal{H}_A)$.

Write the singular value decomposition $|\psi\rangle = \sum_\mu \sqrt{\lambda_\mu}|v_\mu\rangle \otimes |w_\mu\rangle$ with $|v_\mu\rangle \in \mathcal{H}_A'$, $|w_\mu\rangle \in \mathcal{H}_B'$, and recall that the vectors have unit length and are orthogonal. Then

$$(\mathrm{Id}_{\mathcal{H}_A} \otimes \mathrm{Proj}_{\mathcal{M}})|\psi\rangle = \sum_\mu \sqrt{\lambda_\mu}|v_\mu\rangle \otimes \mathrm{Proj}_{\mathcal{M}} |w_\mu\rangle.$$

Since $\psi_{BA}'|w_\mu\rangle = |v_\mu\rangle$, we also have

$$(\tilde{\psi}_{BA}(\mathrm{Proj}_{\mathcal{M}}) \otimes \mathrm{Id}_{\mathcal{H}_B})|\psi\rangle = \sum_\mu \sqrt{\lambda_\mu}\tilde{\psi}_{BA}(\mathrm{Proj}_{\mathcal{M}})|v_\mu\rangle \otimes |w_\mu\rangle.$$

Observe that

$$|\,\mathrm{Proj}_{\mathcal{M}} |w_\mu\rangle| = |\tilde{\psi}_{BA}(\mathrm{Proj}_{\mathcal{M}})|v_\mu\rangle|,$$

so we may write

$$\sum_\mu \sqrt{\lambda_\mu}|v_\mu\rangle\otimes \operatorname{Proj}_{\mathcal{M}}|w_\mu\rangle = \sum_\nu \xi_\nu|v_\nu\rangle\otimes|w'_\nu\rangle \quad \text{and}$$

$$\sum_\mu \sqrt{\lambda_\mu}\tilde{\psi}_{BA}(\operatorname{Proj}_{\mathcal{M}})|v_\mu\rangle\otimes|w_\mu\rangle = \sum_\nu \xi_\nu|v'_\nu\rangle\otimes|w_\nu\rangle,$$

where the $|v'_\mu\rangle$ are orthonormal as are the $|w'_\mu\rangle$. Thus there exist $U_A \in \mathbf{U}(\mathcal{H}_A)$, $U_B \in \mathbf{U}(\mathcal{H}_B)$ such that

$$U_A\otimes U_B\left(\sum_\mu \sqrt{\lambda_\mu}|v_\mu\rangle\otimes \operatorname{Proj}_{\mathcal{M}}|w_\mu\rangle\right) = \sum_\nu \sqrt{\lambda_\nu}\tilde{\psi}_{BA}(\operatorname{Proj}_{\mathcal{M}})|v_\nu\rangle\otimes|w_\nu\rangle.$$

This means we may obtain Bob's $\operatorname{Proj}_{\mathcal{M}}$ by Alice's projection operator $\tilde{\psi}_{BA}(\operatorname{Proj}_{\mathcal{M}})$, followed by isometries U_A and U_B. Thus we may get rid of all of Bob's measurements and Bob's communication with Alice (since she has the result!). Finally, Alice's local actions commute with the local actions of Bob, so we can just combine all of Alice's local actions (unitaries plus measurement) into a single measurement. She then sends the result to Bob, who combines all his unitary actions to a single unitary operator. □

Here is the main theorem on LOCC, called *Nielson's theorem*:

Theorem 6.5.18 (Nielson [**Nie99**]). *Let $|\psi\rangle, |\phi\rangle \in \mathcal{H}_A\otimes\mathcal{H}_B = \mathbb{C}^n\otimes\mathbb{C}^n$ be unit vectors. Then $|\psi\rangle$ may be transformed to $|\phi\rangle$ by LOCC if and only if* $\operatorname{singvals}(|\psi\rangle) \prec \operatorname{singvals}(|\phi\rangle)$.

Recall that $\operatorname{singvals}(|\psi\rangle) = \operatorname{spec}(\rho_{\psi A}) = \operatorname{spec}(\rho_{\psi B})$. Thus we only need to keep track of $\rho_{\psi A}$ in the proof.

Proof. Assume ψ can be transformed to ϕ by LOCC, so there exists an orthogonal decomposition $\mathcal{H}_A = \bigoplus_s \mathcal{M}_s$ and unitary operators $U_s \in \mathbf{U}(\mathcal{H}_B)$ that transform $\rho_\psi := |\psi\rangle\langle\psi|$ to $\rho_\phi := |\phi\rangle\langle\phi|$ via the channel (6.5.4). In particular, after Alice's action, her state is $\rho_{\phi A}$ (as Bob's action will not affect Alice's state). This implies that for each s,

$$\operatorname{Proj}_{\mathcal{M}_s}\rho_{\psi A}\operatorname{Proj}^\dagger_{\mathcal{M}_s} = p_s\rho_{\phi A},$$

where $p_s = \Pr(\mathcal{M}_s) = \operatorname{trace}(\operatorname{Proj}_{\mathcal{M}_s}\rho_{\psi,A})$. By the polar decomposition of $\operatorname{Proj}_{\mathcal{M}_s}\sqrt{\rho_{\psi A}}$ (Proposition/Definition 5.4.4), there exist $U_s \in \mathbf{U}(\mathcal{H}_A)$ such that

$$\begin{aligned}\operatorname{Proj}_{\mathcal{M}_s}\sqrt{\rho_{\psi A}} &= \sqrt{(\operatorname{Proj}_{\mathcal{M}_s}\sqrt{\rho_{\psi A}})(\operatorname{Proj}_{\mathcal{M}_s}\sqrt{\rho_{\psi A}})^\dagger}\,U_s\\ &= \sqrt{p_s\rho_{\phi A}}\,U_s,\end{aligned}$$

and similarly $\sqrt{\rho_{\psi A}}\,\mathrm{Proj}_{\mathcal{M}_s}^{\dagger} = U_s^{-1}\sqrt{p_s \rho_{\phi A}}$. Combining these gives

$$\sqrt{\rho_{\psi A}}\,\mathrm{Proj}_{\mathcal{M}_s}^{\dagger}\,\mathrm{Proj}_{\mathcal{M}_s}\sqrt{\rho_{\psi A}} = p_s U_s^{-1}\rho_{\phi A} U_s.$$

Now sum over s, the projections sum to the identity, and we obtain

$$\rho_{\psi A} = \sum_s p_s U_s^{-1}\rho_{\phi A} U_s,$$

which implies $\mathrm{spec}(\rho_{\psi A}) \prec \mathrm{spec}(\rho_{\phi A})$. If ρ_ψ is invertible, the argument can be run in the other direction to get the reverse conclusion. If not, one splits $\mathcal{H}_A, \mathcal{H}_B$ as in the proof of Proposition 6.5.17 and the argument still goes through. □

Exercise 6.5.19 (Entanglement catalysis). Say $\mathcal{H}_A, \mathcal{H}_B$ are four dimensional and Alice and Bob share $|\psi\rangle = \sqrt{.4}(|00\rangle + |11\rangle) + \sqrt{.1}(|22\rangle + |33\rangle)$. Show that $|\psi\rangle$ cannot be transformed to $|\phi\rangle = \sqrt{.5}|00\rangle + \sqrt{.25}|11\rangle + \sqrt{.25}|22\rangle$ via LOCC. But now say a bank is willing to loan them $|c\rangle = \sqrt{.6}|00\rangle + \sqrt{.4}|11\rangle$. Show that $|\psi\rangle\otimes|c\rangle$ can be transformed by LOCC to $|\phi\rangle\otimes|c\rangle$, so they can obtain $|\phi\rangle$ and return $|c\rangle$ to the bank. In this context, $|c\rangle$ is called a "catalyst".

6.6. Entanglement distillation and dilution

Nielson's theorem 6.5.18 gives us a way to say if ψ is more entangled than ϕ in some cases. In particular, they must lie in the same Hilbert space. We would like to be able to compare the entanglement of pure states in general. We are primarily interested in their asymptotic utility.

To compare the entanglement resources of two states $|\phi\rangle$ and $|\psi\rangle$, we will consider $|\phi\rangle^{\otimes m}$ for large m and determine the largest $d = d(m)$ such that $|\phi\rangle^{\otimes m}$ may be degenerated to $|\psi\rangle^{\otimes d}$ via LOCC. Due to the approximate and probabilistic nature of quantum computing, we will be content to degenerate $|\phi\rangle^{\otimes m}$ to a state that has high fidelity with $|\psi\rangle^{\otimes d}$. We will compare them by comparing each to our standard unit of entanglement $|\mathrm{epr}\rangle$.

Definition 6.6.1. The *entanglement cost* $E_C(\psi)$ of ψ is $\inf_m \frac{d(m)}{m}$ where $|\psi\rangle^{\otimes d(m)}$ can be approximately constructed from $|\mathrm{epr}\rangle^{\otimes m}$ by LOCC with fidelity going to 1 as $m \to \infty$.

The *distillable entanglement* $E_D(\psi)$ of ψ is $\sup_m \frac{\delta(m)}{m}$ where $|\mathrm{epr}\rangle^{\otimes\delta(m)}$ can be approximately constructed from $|\psi\rangle^{\otimes m}$ by LOCC with fidelity going to 1 as $m \to \infty$.

In terms of currency exchange, the entanglement cost should be thought of as the "cost", and the distillable entanglement the "value" of a state.

Entanglement cannot be created by LOCC, so $E_D(\psi) \leq E_C(\psi)$. (Otherwise by going through rounds of LOCC, one could construct an arbitrary number of EPR states.) We will see that for pure states, cost equals value.

Say $|\psi\rangle = \sqrt{p_1}|1\rangle|1\rangle + \cdots + \sqrt{p_n}|n\rangle|n\rangle \in \mathcal{H}_A \otimes \mathcal{H}_B$. Consider

$$|\psi\rangle^{\otimes d} = \sum \sqrt{p_{i_1} \cdots p_{i_d}} |i_1 \cdots i_d\rangle \otimes |i_1 \cdots i_d\rangle.$$

Project $|\psi\rangle^{\otimes d}$ to the ϵ-typical subspace for some small ϵ. Let $|\phi_d\rangle$ denote the image. Recall that the ϵ-typical subspace has dimension at most $2^{d(H(|\psi\rangle\langle\psi|)+\epsilon)}$ by Theorem 6.1.4. We want to show that if Alice and Bob have a pre-shared $|\text{epr}\rangle^{\otimes d(H(|\psi\rangle\langle\psi|)+\epsilon)}$, using LOCC they may obtain a shared

$$|\phi_d\rangle \in (\mathbb{C}^2)^{\otimes d(H(|\psi\rangle\langle\psi|)+\epsilon)} \otimes (\mathbb{C}^2)^{\otimes d(H(|\psi\rangle\langle\psi|)+\epsilon)},$$

where Alice is in possession of the first factor and Bob the second.

Alice is allowed to act locally. Say her space is

$$(\mathbb{C}^2)^{\otimes d(H(|\psi\rangle\langle\psi|)+\epsilon)} \otimes (\mathbb{C}^2)^{\otimes d(H(|\psi\rangle\langle\psi|)+\epsilon)} \otimes (\mathbb{C}^2)^{\otimes d(H(|\psi\rangle\langle\psi|)+\epsilon)},$$

where the last factor is the home of her part of $|\text{epr}\rangle^{\otimes d(H(|\psi\rangle\langle\psi|)+\epsilon)}$. By local actions, she creates $|\phi_d\rangle$ in her first two factors. She then uses the pre-shared $|\text{epr}\rangle^{\otimes d(H(|\psi\rangle\langle\psi|)+\epsilon)}$ and LOCC to teleport the part of $|\phi_d\rangle$ in the second factor to Bob as in §2.5.

Thus $|\psi\rangle^{\otimes d}$ can be constructed from $|\text{epr}\rangle^{\otimes d(H(|\psi\rangle\langle\psi|)+\epsilon)}$ by LOCC with probability of error at most δ using $2^{d(H(|\psi\rangle\langle\psi|)+\epsilon)}$ classical bits and a pre-shared $|\text{epr}\rangle^{\otimes d(H(|\psi\rangle\langle\psi|)+\epsilon)}$. Since we can make ϵ, δ as small as we want, we conclude $E_C(\psi) \leq H(|\psi\rangle\langle\psi|)$.

Now for E_D: say Alice and Bob share $|\psi\rangle^{\otimes m} \in \mathcal{H}_A^{\otimes m} \otimes \mathcal{H}_B^{\otimes m}$ and they want to reliably construct $|\text{epr}\rangle^{\otimes d}$ for $d = d(m)$ as large as possible. For simplicity, assume $\dim \mathcal{H}_A = \dim \mathcal{H}_B = 2$. (Without the assumption, one maps to a space whose dimension is a power of two as in the Quantum Noiseless Channel Theorem (Theorem 6.1.7).) Project $|\psi\rangle^{\otimes m}$ onto the ϵ-typical subspace. The largest coefficient is $2^{-m(H(|\psi\rangle\langle\psi|)-\epsilon)}$. Recall that after projection we have to divide the image by its length to obtain a vector of length one. Thus the largest coefficient of the renormalized projection grows at most by a factor of $\frac{1}{\sqrt{1-\delta}}$. Take d as large as possible such that one still has

$$2^{-d} \geq \frac{2^{-m(H(|\psi\rangle\langle\psi|)-\epsilon)}}{1-\delta}, \tag{6.6.1}$$

i.e., d a littler smaller than $mH(|\psi\rangle\langle\psi|)$. Then

$$\text{spec}((|\psi\rangle\langle\psi|)_{\epsilon-\text{typ}}^{\otimes m}) \prec (2^{-d}, \ldots, 2^{-d}) = \text{spec}((|\text{epr}\rangle\langle\text{epr}|)^{\otimes d}).$$

Thus $(|\psi\rangle\langle\psi|)_{\epsilon-\text{typ}}^{\otimes m}$ may be transformed to $(|\text{epr}\rangle\langle\text{epr}|)^{\otimes d}$ by LOCC. We conclude $E_D(\psi) \geq H(|\psi\rangle\langle\psi|)$.

Together the inequalities give

$$H(|\psi\rangle\langle\psi|) \leq E_D(\psi) \leq E_C(\psi) \leq H(|\psi\rangle\langle\psi|),$$

and we see they are all equalities.

In summary:

Theorem 6.6.2 ([**BBP⁺96**]). *For a pure state* $|\psi\rangle \in \mathbb{C}^n \otimes \mathbb{C}^n$,

$$E_D(\psi) = E_C(\psi).$$

Remark 6.6.3. In classical computation one can reproduce information, but this cannot be done with quantum information. This is because the map $|\psi\rangle \mapsto |\psi\rangle\otimes|\psi\rangle$, called the *Veronese map* in algebraic geometry, is not a linear map. This observation is called the *no cloning theorem* in the quantum literature. One can define a linear map, e.g., $\mathbb{C}^2 \to \mathbb{C}^2\otimes\mathbb{C}^2$ that duplicates basis vectors; i.e., $|0\rangle \mapsto |0\rangle\otimes|0\rangle$ and $|1\rangle \mapsto |1\rangle\otimes|1\rangle$. But then $\alpha|0\rangle + \beta|1\rangle \mapsto \alpha|0\rangle\otimes|0\rangle + \beta|1\rangle\otimes|1\rangle \neq (\alpha|0\rangle + \beta|1\rangle)^{\otimes 2}$.

6.6.1. Cost and value of mixed states.

For mixed states $\rho \in \mathcal{D}(\mathcal{H}_A\otimes\mathcal{H}_B)$, one may still define $E_C(\rho)$ and $E_D(\rho)$, but there are examples where they differ. As of this writing there is no canonical measure of entanglement for density operators.

Example 6.6.4. Let $\dim \mathcal{H}_A = \dim \mathcal{H}_B = 3$. Write

$$|\psi_{ij}\rangle = \frac{1}{\sqrt{2}}(|i\rangle\otimes|j\rangle - |j\rangle\otimes|i\rangle) \in \mathcal{H}_A\otimes\mathcal{H}_B$$

and consider

$$\rho := \frac{1}{3}(|\psi_{23}\rangle\langle\psi_{23}| + |\psi_{31}\rangle\langle\psi_{31}| + |\psi_{12}\rangle\langle\psi_{12}|) \in \mathcal{D}(\mathcal{H}_A\otimes\mathcal{H}_B).$$

In [**Yur03**], it is shown that $E_C(\rho) = 1$ (the calculation is long, but straightforward) but $E_D(\rho) < 1$.

For a general density operator, one may also define the *entanglement of formation* to be

$$E_F(\rho) := \inf \sum_j p_j H(\text{trace}_{\mathcal{H}_B}(|\mu_j\rangle\langle\mu_j|))$$

where the infimum is over all decompositions $\rho = \sum_i p_i|\mu_i\rangle\langle\mu_i|$, where the μ_j are orthonormal unit vectors but we do *not* require them to be eigenvectors of ρ. A desirable property of an entanglement measure is additivity. Unfortunately E_F fails to be additive.

In Example 6.6.4, $E_F(\rho) = E_C(\rho) = 1$.

In [**Yur03**], it is shown that $E_C(\rho) = \lim_{d\to\infty} \frac{1}{d} E_F(\rho^{\otimes d})$.

Let's make a wish list of what we might want from an entanglement measure E on density operators:

- Nonincreasing under LOCC.
- If ρ is a product state, i.e., $\rho = \rho_A \otimes \rho_B$, then $E(\rho) = 0$.

The two conditions together imply any state constructible from a product state by LOCC should also have zero entanglement. Hence the following definition:

Definition 6.6.5. A density operator $\rho \in \mathcal{D}(\mathcal{H}_1 \otimes \cdots \otimes \mathcal{H}_N)$ is *separable* if $\rho = \sum_i p_i \rho_{i,1} \otimes \cdots \otimes \rho_{i,N}$, where $\rho_{i,\alpha} \in \mathcal{D}(\mathcal{H}_\alpha)$ and $\bar{p} = (p_1, \dots, p_m)$ is a probability distribution. If ρ is not separable, one says ρ is *entangled*.

The simplest example of an entangled density operator is the density operator associated to an epr state.

So we replace our second condition by requiring that $E(\rho) = 0$ for any separable ρ.

Definition 6.6.6. An *entanglement monotone E* is a function on density operators on $\mathcal{H}_A \otimes \mathcal{H}_B$ that is nonincreasing under LOCC.

This implies E is zero on separable states. One would also like E to agree with E_C, E_D on pure states.

Recall the mutual information $I(\rho_A : \rho_B) := H(\rho_A) + H(\rho_B) - H(\rho_{AB})$. This vanishes on product states.

If ρ is pure, then $I(\rho_A : \rho_B) = 2H(\rho_A) = 2H(\rho_B)$, so a first idea for an entanglement monotone on density operators would be to take half the mutual information.

However, the mutual information fails to be zero on separable states. To fix this, define the *squashed entanglement* [**CW04**]:

$$E_{sq}(\rho_{AB}) := \inf_C \left\{ \frac{1}{2} I(\rho_A : \rho_B | \rho_C) \mid \rho_{AB} = \operatorname{trace}_{\mathcal{H}_C}(\rho_{ABC}) \right\}.$$

Exercise 6.6.7. Show that E_{sq} vanishes on separable states.

According to [**CW04**], the name was chosen because "the right choice of conditioning system reduces the quantum mutual information between $\mathcal{H}_A$ and $\mathcal{H}_B$, thus 'squashing out' the nonquantum correlations."

The squashed entanglement is (at least as of this writing) hard to compute in general, but it does have the desirable property that it is additive under tensor products:

$$E_{sq}(\rho_{AB} \otimes \rho_{CD}) = E_{sq}(\rho_{AB}) + E_{sq}(\rho_{CD}).$$

At this writing, determining entanglement measures for general density operators is an active area of research. See [**HHHH09**] for a discussion of the state of the art up to 2008. Despite there being numerous articles on the subject, it does not appear to have fundamentally advanced since then. See, e.g., [**VBHAF23**], where they state "to our best knowledge, a directly or numerically computable measure for the entanglement of multipartite mixed states is still lacking."

Chapter 7

Representation theory and quantum information

Representation theory is the systematic use of symmetry in linear algebra. In quantum information theory, we have so far seen relations between a state ρ_{AB} on $\mathcal{H}_A \otimes \mathcal{H}_B$ and its partial traces ρ_A, ρ_B. Representation theory allows one to make deeper connections between these objects, as well as between states obtained from a density operator on tensor products of larger numbers of spaces. In particular, we outline a representation-theoretic proof of strong subadditivity that does not rely on convexity.

We begin, in §7.1, with a crash course in representation theory focusing on Schur-Weyl duality. This material is covered in numerous sources. We are brief and refer the reader to any of [**FH91, Pro07, Lan12**] for more details. In §7.2, Schur-Weyl duality is used to obtain a representation-theoretic interpretation of the quantum ϵ-typical subspaces. The quantum marginal problem addressed by Klayatchko and independently by Christandl and Mitchison is discussed in §7.3. We present the proofs from [**CHM07**]. A generalization of the quantum marginal results to arbitrary density operators on tripartite states from [**CcW18**] is presented in §7.4, where the representation-theoretic proof of strong subadditivity is outlined. For readers familiar with basic definitions from commutative algebra and representation theory, §7.5 contains a proof of the finite generation of the semi-group $\mathrm{Kron}_{m,n,k}$ consisting of nonvanishing Kronecker coefficients, and it contains a monotonicity result of L. Manivel on the growth of Kronecker coefficients.

7.1. Representation theory

Informally, a representation of a group G is a realization of G as a subgroup of the group of invertible $m \times m$-matrices for some m.

7.1.1. Definitions. We will be primarily concerned with the representation theory of the permutation group $\mathfrak{S}_d$ (also called the *symmetric group*), the general linear group GL_n, and unitary group $\mathbf{U}(n)$. In order to distinguish the role of a Hermitian inner-product, when the Hermitian inner-product is not used or needed, we generally denote vector spaces by letters V, W and use $\mathcal{H}_A, \mathcal{H}_B$ for Hilbert spaces.

Definition 7.1.1. Let G be a group. A *representation* of G is a group homomorphism $\mu : G \to \mathrm{GL}(V)$ for some vector space V. One says G *acts* on V and that V is a *G-module*.

When the action of G is clear, we sometimes simply denote it with a dot "$\cdot$", or omit μ from the notation altogether.

Definition 7.1.2. Let V be a G-module, and let $W \subset V$ be a subspace. One says W is a *submodule* if for all $g \in G$ and $w \in W$, $\mu(g)w \in W$. The representation V is *irreducible* if it has no proper submodules.

For example, the action of $\mathfrak{S}_d$ on $\mathbb{C}^d$ given by permuting basis elements is not irreducible:

Exercise 7.1.3. Write $\mathbb{C}^d = W_1 \oplus W_2$ where W_1, W_2 are $\mathfrak{S}_d$-submodules. ⊚

Definition 7.1.4. A G-module V is *trivial* if $\mu(g)v = v$ for all $g \in G$ and $v \in V$.

Definition 7.1.5. Let V, W be G-modules. A linear map $f : V \to W$ is a *G-module map* if for all $v \in V$ and $g \in G$, $f(\mu_V(g)v) = \mu_W(g)f(v)$. Let $\mathrm{Hom}_G(V, W) \subset \mathrm{Hom}(V, W)$ denote the vector space of G-module maps $V \to W$. The modules V, W are said to be *isomorphic* if there exists a G-module map between them that is a linear isomorphism.

We will identify representations up to isomorphism. For example, when we write "we construct all the irreducible $\mathfrak{S}_d$-modules", we mean we construct a representative of each isomorphism class.

Lemma 7.1.6 (Schur's lemma). *Let V, W be irreducible finite-dimensional complex G-modules, and let $f : V \to W$ be a G-module map. Then either $f = 0$ or f is an isomorphism. If furthermore $V = W$, then $f = \lambda\,\mathrm{Id}$ for some constant λ.*

Exercise 7.1.7. Prove Schur's lemma. ⊚

If G acts on V by μ_V, it acts on V^* by μ_{V^*}, which is defined by the equation

$$[\mu_{V^*}(g)(\alpha)](v) := \alpha(\mu_V(g)^{-1}v)$$

for all $\alpha \in V^*$, $v \in V$, $g \in G$. If G acts on V_1, V_2, by μ_1, μ_2, then it acts on $V_1 \otimes V_2$ by

$$\mu(g)(v_1 \otimes v_2) = \mu_1(g)v_1 \otimes \mu_2(g)v_2.$$

In particular, if G acts on V, it acts on all tensor powers of V and V^*. These actions are called *induced actions*.

Exercise 7.1.8. Show that if G acts on V, then the induced action on $\mathrm{End}(V) = V \otimes V^*$ contains a trivial submodule. Show that moreover if V is irreducible, the trivial submodule is unique. ⊚

7.1.2. Representations of the permutation group and Schur-Weyl duality.

We describe the irreducible modules for the permutation group. For a proof, see, e.g., [**Lan17**, §8.6.8], [**FH91**, §I.4], or [**Mac95**, §I.7].

Definition 7.1.9. A *partition* π of d is a d-tuple of nonnegative integers $\pi = (\pi_1, \dots, \pi_d)$, $\pi_i \geq \pi_{i+1} \geq 0$ and $\sum_i \pi_i = d$. Introduce the notation $|\pi| := d$. Let $\ell(\pi)$ denote the number of nonzero π_j, the *length* of π.

Theorem 7.1.10. *The irreducible representations of $\mathfrak{S}_d$ are in one-to-one correspondence with the partitions of d.*

Given a partition π, let $[\pi]$ denote the (isomorphism class of the) corresponding irreducible $\mathfrak{S}_d$-module.

What follows is a recipe for constructing all the irreducible $\mathfrak{S}_d$-modules as submodules of $V^{\otimes d}$, where $\dim V \geq d$. It will be clear from the construction that we obtain $\mathfrak{S}_d$-modules, but we rely on Theorem 7.1.10 to assert they are irreducible and are all the irreducible $\mathfrak{S}_d$-modules. As a bonus, we will also construct the irreducible $\mathrm{GL}(V)$-modules appearing in $V^{\otimes d}$.

The decomposition of a Hilbert space $\mathcal{H}^{\otimes d}$ as a module for the unitary group is the same as the decomposition as a $\mathrm{GL}(\mathcal{H})$-module. We will discuss the decomposition in terms of $\mathrm{GL}(V)$ and only introduce the unitary group when it becomes relevant.

Given a vector space V, $\mathfrak{S}_d$ acts on $V^{\otimes d}$ by permuting the factors:

$$\mu(\sigma)(v_1 \otimes \cdots \otimes v_d) := v_{\sigma^{-1}(1)} \otimes \cdots \otimes v_{\sigma^{-1}(d)}.$$

One takes the inverse so that $\mu(\sigma)\mu(\tau)(v_1 \otimes \cdots \otimes v_d) = \mu(\sigma\tau)(v_1 \otimes \cdots \otimes v_d)$. This action commutes with the action of each $g \in \mathrm{GL}(V)$, where

$$g \cdot (v_1 \otimes \cdots \otimes v_d) = (gv_1) \otimes \cdots \otimes (gv_d).$$

In other words, for $\sigma \in \mathfrak{S}_d$, $\mu(\sigma) : V^{\otimes d} \to V^{\otimes d}$ is a $\mathrm{GL}(V)$-module map. Similarly, for all $g \in \mathrm{GL}(V)$, one has $g\cdot : V^{\otimes d} \to V^{\otimes d}$ is an $\mathfrak{S}_d$-module map.

To visualize a partition $\pi = (\pi_1, \dots, \pi_d)$, define the *Young diagram* associated to π, denoted $\mathrm{YD}(\pi)$, to be a collection of left-aligned boxes with π_j boxes in the j-th row, as in Figure 7.1.1.

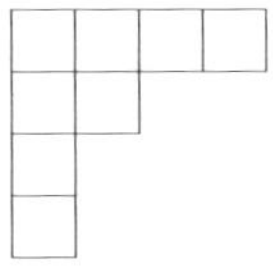

Figure 7.1.1. Young diagram for $\pi = (4, 2, 1, 1)$

In the notation of Definition 7.1.9, $\ell(\pi)$ is the number of boxes in the first column of the Young diagram of π.

Label the boxes in the diagram by $\{1, \dots, d\}$; such is called a *Young tableau without repetitions*. For example:

$$\begin{array}{|c|c|c|c|} \hline 3 & 7 & 2 & 5 \\ \hline 1 & 4 \\ \cline{1-2} 6 \\ \cline{1-1} 8 \\ \cline{1-1} \end{array}$$

The *default Young tableau* is labeled left to right and top to bottom:

$$\begin{array}{|c|c|c|c|} \hline 1 & 2 & 3 & 4 \\ \hline 5 & 6 \\ \cline{1-2} 7 \\ \cline{1-1} 8 \\ \cline{1-1} \end{array}$$

Consider the following $\mathrm{GL}(V)$-module map $V^{\otimes d} \to V^{\otimes d}$:

$$v_1 \otimes \cdots \otimes v_d \mapsto \sum_{\sigma \in \mathfrak{S}_d} v_{\sigma(1)} \otimes \cdots \otimes v_{\sigma(d)},$$

which symmetrizes $v_1 \otimes \cdots \otimes v_d$. This map is the projection to the irreducible $\mathrm{GL}(V)$-submodule $S^d V$. In this chapter we use the word "projection" in a more general sense than previously: we only require the image of an endomorphism f to be equal to the image of the map f^2.

More generally, given a Young tableau without repetitions, consider the following projection operator on $V^{\otimes d}$ constructed from it: first write $V^{\otimes d} = V_1 \otimes V_2 \otimes \cdots \otimes V_d$, where the subscript is to remember the position. Then for each row in the Young tableau, symmetrize the corresponding copies of V; e.g., for $\pi = (4, 2, 2, 1)$, after the symmetrization, one obtains an element of $S^4 V \otimes S^2 V \otimes S^2 V \otimes V \subset V^{\otimes 8}$. Next, skew-symmetrize along the columns. Explicitly, introduce the notation $\mathfrak{S}[j_1, \dots, j_k]$ for the permutation group on the indices $\{j_1, \dots, j_k\}$. Then the symmetrization map for $\pi = (4, 2, 2, 1)$ with the default Young tableau is

$$v_1 \otimes \cdots \otimes v_9 \mapsto \sum_{\substack{\sigma_1 \in \mathfrak{S}[1,2,3,4] \\ \sigma_2 \in \mathfrak{S}[5,6], \sigma_3 \in \mathfrak{S}[7,8]}} v_{\sigma_1(1)} \otimes v_{\sigma_1(2)} \otimes v_{\sigma_1(3)} \otimes v_{\sigma_1(4)} \otimes v_{\sigma_2(5)} \otimes v_{\sigma_2(6)} \otimes v_{\sigma_3(7)} \otimes v_{\sigma_3(8)} \otimes v_9$$

and the skew-symmetrization map is

$$w_1 \otimes \cdots \otimes w_9 \mapsto \sum_{\substack{\sigma \in \mathfrak{S}[1,5,7,9] \\ \tau \in \mathfrak{S}[2,6,8]}} \operatorname{sgn}(\sigma) \operatorname{sgn}(\tau) w_{\sigma(1)} \otimes w_{\tau(2)} \otimes w_3 \otimes w_4 \otimes w_{\sigma(5)} \otimes w_{\tau(6)} \otimes w_{\sigma(7)} \otimes w_{\tau(8)} \otimes w_{\sigma(9)}.$$

For another example, if we take the default Young tableau for $\pi = (2,1)$, the maps are

$$u \otimes v \otimes w \mapsto u \otimes v \otimes w + v \otimes u \otimes w$$

followed by

$$u \otimes v \otimes w + v \otimes u \otimes w \mapsto u \otimes v \otimes w - w \otimes v \otimes u + v \otimes u \otimes w - w \otimes u \otimes v.$$

For the default Young tableau, write the resulting map as

$$\mathrm{Proj}_{\pi-\mathrm{def}} : V^{\otimes d} \to V^{\otimes d},$$

and let $S_{\pi-\mathrm{def}}V := \mathrm{Proj}_{\pi-\mathrm{def}}(V^{\otimes d})$ denote the image.

Exercise 7.1.11. Write down the map $\mathrm{Proj}_{(2,2)-\mathrm{def}} : V^{\otimes 4} \to V^{\otimes 4}$. ⊚

Exercise 7.1.12. Show that if $d > \dim V$, then it is possible that $\mathrm{Proj}_{\pi-\mathrm{def}}(V^{\otimes d}) = 0$ but that otherwise it is always nonzero. ⊚

Since $\mathrm{Proj}_{\pi-\mathrm{def}}$ is a composition of actions by elements of $\mathfrak{S}_d$, it is a $\mathrm{GL}(V)$-module map, and thus $S_{\pi-\mathrm{def}}V$ is a $\mathrm{GL}(V)$-submodule.

Fact. When nonzero, $S_{\pi-\mathrm{def}}V$ is an irreducible $\mathrm{GL}(V)$-module, and if $\pi \neq \nu$, then $S_{\pi-\mathrm{def}}V$ is not isomorphic to $S_{\nu-\mathrm{def}}V$. (This is a consequence of Schur-Weyl duality and the theorem of the highest weight discussed below.)

Let $V = \mathbb{C}^N$ be equipped with its standard basis, which in this chapter will be denoted $e_1, \dots, e_N$, and assume $N \geq d$. Set

$$v_{\pi-\mathrm{def}} := \mathrm{Proj}_{\pi-\mathrm{def}}(e_1^{\otimes \pi_1} \otimes e_2^{\otimes \pi_2} \otimes \cdots \otimes e_d^{\otimes \pi_d}).$$

For any group G and any G-module W, the span of the G-orbit of any $v \in W$ is a submodule (or all of W).

Fact. The span of the $\mathfrak{S}_d$-orbit of $v_{\pi-\mathrm{def}}$ is an irreducible $\mathfrak{S}_d$-module isomorphic to $[\pi]$. (This is a consequence of Schur-Weyl duality discussed below.)

Example 7.1.13. The partitions $\pi = (d, 0, \ldots, 0) =: (d)$, $\pi = (1, \ldots, 1) =: (1^d)$, and $\pi = (d-1, 1)$, respectively, correspond to the one-dimensional trivial representation, one-dimensional sign representation $\mu(\sigma)v = \operatorname{sgn}(\sigma)v$, and the action on $\mathbb{C}^{d-1} \subset \mathbb{C}^d$ where $\mathbb{C}^{d-1}$ is the subspace of vectors whose entries add to zero.

If one defines the corresponding map for a different Young tableau without repetitions associated to π, one obtains a $\mathrm{GL}(V)$-module isomorphic to $S_{\pi-\mathrm{def}}V$ because one can map from one to the other via a composition of elements of $\mathfrak{S}_d$, and it is straightforward to check that the maps are bijective. Let $S_\pi V$ denote the isomorphism class of $S_{\pi-\mathrm{def}}V$, i.e., the equivalence class of $\mathrm{GL}(V)$-modules admitting a $\mathrm{GL}(V)$-isomorphism with $S_{\pi-\mathrm{def}}V$.

Definition 7.1.14. If W, M are G-modules with M irreducible, the *isotypic component* of M in W is the largest submodule of W isomorphic to $M^{\oplus m}$ for some m. The integer m is called the *multiplicity* of M in W. One writes $\operatorname{mult}(M, W) = m$. Let W^G denote the isotypic component of the trivial representation in W, called the space of *invariants* under the action of G.

Exercise 7.1.15. Show that $(V^{\otimes d})^{\mathfrak{S}_d} = S^d V$.

Exercise 7.1.16. Show that $\operatorname{mult}(M, W) = \dim \operatorname{Hom}_G(M, W)$.

Theorem 7.1.17 (Schur-Weyl duality). *As a* $\mathrm{GL}(V) \times \mathfrak{S}_d$*-module,*

$$V^{\otimes d} = \bigoplus_{\substack{|\pi|=d \\ \ell(\pi)\leq \dim V}} S_\pi V \otimes [\pi].$$

Note that this decomposition into irreducible $\mathrm{GL}(V) \times \mathfrak{S}_d$-submodules is multiplicity free; i.e., each irreducible $\mathrm{GL}(V) \times \mathfrak{S}_d$-module appearing occurs with multiplicity one.

For a proof, see, e.g., [**Lan17**, Thm. 8.7.1.2] or [**GW09**, Chap. 4].

Theorem 7.1.17 implies

$$\operatorname{mult}([\pi], V^{\otimes d}) = \dim S_\pi V \quad \text{and} \quad \operatorname{mult}(S_\pi V, V^{\otimes d}) = \dim[\pi].$$

7.1.3. Dimensions of GL_m-modules and $\mathfrak{S}_d$-modules.

We will be concerned not just with the dimension of individual modules, but the asymptotics of their growth, where "growth" is made precise in Exercise 7.1.21 and Lemma 7.1.24 below. For a box x in the Young diagram of π, define the *hook length* $h(x)$ of x to be the number of boxes to the right of it in its row plus the number of boxes below it in its column plus one. Define the *content* $c(x)$ of x to be zero if x is on the main diagonal, j if it is on the j-th diagonal above the main diagonal, and $-j$ if it is on the j-th diagonal below the main diagonal. For example, we have the following hook lengths and contents for (4, 2, 2, 1):

7	5	2	1
4	2		
3	1		
1			

and

0	1	2	3
−1	0		
−2	−1		
−3			

One has the following dimension formulas where $|\pi| = d$:

$$\dim S_\pi \mathbb{C}^m = \prod_{x\in\pi} \frac{m + c(x)}{h(x)} \tag{7.1.1}$$

$$= \prod_{1\leq i<j\leq m} \frac{\pi_i - \pi_j + j - 1}{j - i}, \text{ and} \tag{7.1.2}$$

$$\dim[\pi] = \frac{d!}{\prod_{x\in\pi} h(x)}. \tag{7.1.3}$$

See, e.g., [**FH91**, §4.12, Thm. 6.3, Ex. 6.4] or [**Pro07**, §6.2, §9.5] for proofs.

Exercise 7.1.18. Let $\bar{p}_m(d)$ denote the number of partitions of d with at most m parts. Show that

$$\bar{p}_m(d) = \sum_{i=0}^{\lfloor \frac{d}{m} \rfloor} \bar{p}_{m-1}(d - im).$$

Exercise 7.1.19. Use Exercise 7.1.18 and induction to show that when $d, m > 1$, there are less than d^m partitions of d with at most m parts. (With a little more work you can show $\bar{p}_m(d) = \frac{1}{m!(m-1)!} d^{m-1} + O(d^{m-2})$.)

Aside 7.1.20. There are $O(\frac{1}{d} e^{\pi\sqrt{\frac{2d}{3}}})$ partitions of d [**HR00**], but we are interested in partitions fitting inside a $d \times m$ box.

Exercise 7.1.21. Show that

$$\dim S_\pi \mathbb{C}^m \leq (d+1)^{\binom{m}{2}}. \tag{7.1.4}$$ ⊙

Exercise 7.1.21 shows that if one fixes m and lets $d \to \infty$, for any sequence of partitions $\pi(d)$, with $|\pi(d)| = d$, the dimension of $S_{\pi(d)}\mathbb{C}^m$ is bounded by a polynomial in d.

However, $\dim(\mathbb{C}^m)^{\otimes d} = m^d$ is exponential in d. Write

$$m^d = \sum_{|\pi|=d, \ell(\pi)\leq m} (\dim S_\pi \mathbb{C}^m)(\dim[\pi]).$$

By Exercise 7.1.19 there are poly(d) summands, and by Exercise 7.1.21 the contribution of the dimension of the GL_m-module to each summand is poly(d), giving rise to the following:

Fundamental observation. In the decomposition of $V^{\otimes d}$, when d is large the $\dim[\pi]$'s account for most of the dimension; in particular, at least one of the $[\pi]$'s must be of dimension exponential in d.

Remark 7.1.22. Despite (7.1.4) being a coarse estimate, S. Mkrtchyan [**Lan19**, Prop. 7.1.4.3] showed that the actual largest dimension as a function of d is within a constant factor of (7.1.4).

Next we estimate how $\dim[\pi]$ "grows with d".

Exercise 7.1.23. Show that

$$\Pi_{j=1}^m \pi_j! \leq \Pi_{x\in YD(\pi)} h(x) \leq \Pi_{j=1}^m (\pi_j + m - j)!\,. \quad ⊚$$

Thus

$$\frac{d!}{\Pi_{j=1}^m(\pi_j + m - j)!} \leq \dim[\pi] \leq \frac{d!}{\Pi_{j=1}^m \pi_j!} = \binom{d}{\pi_1,\dots,\pi_m}. \tag{7.1.5}$$

By (4.2.6)

$$\log\binom{d}{\pi_1,\dots,\pi_m} = dH\left(\frac{\pi_1}{d},\dots,\frac{\pi_m}{d}\right) + O(\log(d)).$$

For a partition $\pi = (\pi_1,\dots,\pi_d)$ of d, introduce the notation $\bar{\pi} := (\frac{\pi_1}{d},\dots,\frac{\pi_d}{d})$ which is a probability distribution on $\{1,\dots,d\}$.

Thus when $|\pi|$ is sufficiently large, using that

$$(\pi_j + m - j)! = \pi_j! \prod_{u=0}^{m-j-1} (\pi_j + m - j - u),$$

we obtain

$$\begin{aligned} dH(\bar{\pi}) - O(\log(d)) - \sum_{j=1}^{\ell(\pi)} \log\left(\prod_{s=1}^{m-j}(\pi_j + s)\right) & \\ \leq \log\dim[\pi] & \\ \leq dH(\bar{\pi}) + O(\log(d)). & \end{aligned} \tag{7.1.6}$$

Exponentiating, we obtain

$$\frac{2^{|\pi|H(\overline{\pi}) - O(\log(d))}}{(|\pi|+1)^{\ell(\pi)^2}} \leq \dim[\pi] \leq 2^{|\pi|H(\overline{\pi}) + O(\log(d))}. \tag{7.1.7}$$

For $\pi = (\pi_1, \dots, \pi_m)$ write

$$N\pi := (N\pi_1, \dots, N\pi_m),$$

which is a partition of $N|\pi|$.

Lemma 7.1.24. *For any partition π, one has*

$$\lim_{N\to\infty} \frac{1}{N|\pi|} \log(\dim[N\pi]) = H(\overline{\pi}).$$

Proof. Equation (7.1.6) says

$$\begin{aligned} N|\pi|H(\bar{\pi}) - O(\log(N|\pi|)) - \ell(\pi)^2 \log(N|\pi| + 1) \\ \le \log\dim[N\pi] \\ \le N|\pi|H(\bar{\pi}) + O(\log(N|\pi|)). \end{aligned}$$

Now divide by $N|\pi|$ and let $N \to \infty$. □

7.2. Schur-Weyl duality and ϵ-typical subspaces

7.2.1. The spectrum estimation theorem. We now describe, for a given density operator ρ, the ϵ-typical and δ-small subspaces of $\mathcal{H}^{\otimes d}$ from §6.1.3 in terms of projections onto isotypic components $S_\pi\mathcal{H}\otimes[\pi]$. Namely, for a density operator ρ on $\mathcal{H}$, the Spectrum Estimation Theorem 7.2.1 states that the projection of $\rho^{\otimes d}$ onto the $\mathrm{GL}(\mathcal{H}) \times \mathfrak{S}_d$ submodule $\mathrm{End}(S_\pi\mathcal{H}\otimes[\pi])$ of $\mathrm{End}(\mathcal{H}^{\otimes d})$ for large d is negligible unless $\bar{\pi}$ is close to $\mathrm{spec}(\rho)$.

Recall the relative entropy from §6.2.1:

$$H(\bar{p}\|\bar{q}) = -\sum_i p_i \log\frac{q_i}{p_i} = -H(\bar{p}) - \sum_i p_i \log(q_i),$$

which may be thought of as measuring how close p, q are because it is nonnegative, and it is zero if and only if $p = q$.

Theorem 7.2.1 (Spectrum Estimation Theorem, [**ARS88, KW01**]). *Let* $\dim\mathcal{H} = n$, *and let $\rho \in \mathcal{D}(\mathcal{H})$ be a density operator. Let $|\pi| = d$, assume $\ell(\pi) \le n$, and in what follows, if $\ell(\pi) < n$, add zeros to it so we may write $\pi = (\pi_1, \dots, \pi_n)$. Let* $\mathrm{Proj}_\pi : \mathcal{H}^{\otimes d} \to S_\pi\mathcal{H}\otimes[\pi]$ *be the projection operator. Write* $\mathrm{spec}(\rho) = (r_1, \dots, r_n)$. *Then*

$$\begin{aligned} \mathrm{trace}(\mathrm{Proj}_\pi \rho^{\otimes d}) &\le (d+1)^{\binom{n}{2}} 2^{-dH(\bar{\pi}\|\,\mathrm{spec}(\rho))} \\ &= (d+1)^{\binom{n}{2}} 2^{dH(\bar{\pi})} 2^{\sum_i \pi_i \log(r_i)}. \end{aligned}$$

Thus, $\mathrm{trace}(\mathrm{Proj}_\pi \rho^{\otimes d})$ is roughly bounded above by $\dim(S_\pi\mathcal{H}\otimes[\pi])$ times $2^{\sum_i \pi_i \log(r_i)} = r_1^{\pi_1} \cdots r_n^{\pi_n}$.

Theorem 7.2.1 is proved in §7.2.3.

We need a few more notions from representation theory for the proof.

7.2.2. Weight spaces and the dominance order. Let $e_1, \dots, e_n$ be a basis of V, and write elements of the induced basis of $V^{\otimes d}$ as $e_I = e_{i_1} \otimes \cdots \otimes e_{i_d}$. In analogy with our definition in Chapter 4, we have:

Definition 7.2.2. The *weight* of e_I is $\mathrm{wt}(e_I) := (w_1, \dots, w_n)$ where w_j is the number of i_t's equal to j. A vector $X \in V^{\otimes d}$ is a *weight vector* if it is a linear combination of basis vectors all having the same weight.

We next determine, for a given partition π, which basis vectors are in the kernel of the projection map

$$\mathrm{Proj}_\pi : V^{\otimes d} \to S_\pi V \otimes [\pi].$$

First consider the case $\pi = (1^d)$, the projection onto $\Lambda^d V$. Apply this to some e_I. If the i_t are not all distinct, e_I will map to zero.

Exercise 7.2.3. Show that if $\ell(\pi) \leq n$ and $n|d$, then $\mathrm{Proj}_\pi(e_1^{\otimes \frac{d}{n}} \otimes \cdots \otimes e_n^{\otimes \frac{d}{n}}) \neq 0$, and show that $\mathrm{Proj}_\pi(e_1^d) = 0$ unless $\pi = (d)$. ⊚

The projection of e_I to $S_\pi V \otimes [\pi]$ involves skew-symmetrizing over $\ell(\pi)$ positions. If there are not at least $\ell(\pi)$ distinct indices in I, the projection of e_I to $S_\pi V \otimes [\pi]$ is zero. But this is not the only skew-symmetrization. Say q_2 is the number of boxes in the second column of the Young diagram of π. Then, after removing $q_1 := \ell(\pi)$ distinct indices from I, there must still be q_2 distinct indices left or the projection is zero. And after removing these $q_1 + q_2$ indices, in order to have a nonzero projection, there must be at least the number of boxes in the third column, call this q_3, distinct indices remaining. For a partition π, let π' denote the partition whose Young diagram is the transpose of the Young diagram of π, so $\pi' = (q_1, q_2, q_3, \dots, q_{\pi_1})$.

Recall the dominance order (Definition 6.5.3) to compare partitions. Continuing the above reasoning, we see that unless $\pi' \prec \mathrm{wt}(e_I)'$, we must have $\mathrm{Proj}_\pi(e_I) = 0$. With a little more work, you will show this necessary condition is also sufficient.

Exercise 7.2.4. Show for partitions μ, ν of d that $\mu \prec \nu$ if and only if $\nu' \prec \mu'$.

Exercise 7.2.5. Show that the projection of e_I to $S_\pi V \otimes [\pi]$ is nonzero if and only if $\mathrm{wt}(e_I) \prec \pi$. ⊚

For example, when $\pi = (d)$ no basis vector maps to zero, and when $\pi = (1, \dots, 1) = (1^d)$, only basis vectors with weights having d of the w_i equal to 1 have nonzero images.

7.2.3. Proof of the spectrum estimation theorem and its consequences for ϵ-typical subspaces.

Proof of Theorem 7.2.1. Write $\bar{r} = \operatorname{spec}(\rho) = (r_1, \dots, r_n)$ with $r_j \geq r_{j+1}$. Choose an orthonormal basis $(e_1, \dots, e_n)$ of $\mathcal{H}$ consisting of eigenvectors of ρ, with r_i the eigenvalue of e_i, so $\rho^{\otimes d} = \sum_I r_I |e_I\rangle\langle e_I|$, where $r_I = r_{i_1} \cdots r_{i_d}$. The eigenvectors that do not project to zero in $[\pi]\otimes S_\pi\mathcal{H}$ are those with $\operatorname{wt}(e_I) \prec \pi$. Note $\operatorname{wt}(e_I) = \pi$ implies $r_I = r_1^{\pi_1} \cdots r_n^{\pi_n}$, and more generally $\operatorname{wt}(e_I) \prec \pi$ implies $r_I \leq r_1^{\pi_1} \cdots r_n^{\pi_n}$. Thus

$$\begin{aligned} \operatorname{Proj}_\pi(\rho^{\otimes d}) &= \sum_{\operatorname{wt}(e_I)\prec\pi} r_I \operatorname{Proj}_\pi(|e_I\rangle\langle e_I|) \\ &\leq r_1^{\pi_1} \cdots r_n^{\pi_n} \sum_{\operatorname{wt}(e_I)\prec\pi} \operatorname{Proj}_\pi(|e_I\rangle\langle e_I|), \end{aligned}$$

and

$$\begin{aligned} \operatorname{trace}(\operatorname{Proj}_\pi(\rho^{\otimes d})) &\leq r_1^{\pi_1} \cdots r_n^{\pi_n} \dim([\pi]\otimes S_\pi\mathcal{H}) \\ &\leq (d+1)^{\binom{n}{2}} 2^{dH(\bar{\pi})} r_1^{\pi_1} \cdots r_n^{\pi_n}. \end{aligned}$$

The last line follows from Exercise 7.1.21 and equation (7.1.7). □

Let SPEC_n denote the set of possible ordered spectra for density operators on $\mathbb{C}^n$; i.e.,

$$\mathrm{SPEC}_n = \left\{ (r_1, \dots, r_n) \mid r_1 \geq r_2 \cdots \geq r_n \geq 0, \sum_i r_i = 1 \right\}.$$

For $r \in \mathrm{SPEC}_n$, let

$$B_\epsilon(r) := \{ r' \in \mathrm{SPEC}_n \mid \|r - r'\|_1 < \epsilon \}$$

and let

$$\mathrm{Rep}_{\epsilon,d,r} := \bigoplus_{\substack{|\lambda|=d,\\ \bar{\lambda}\in B_\epsilon(r)}} [\lambda]\otimes S_\lambda\mathcal{H}.$$

We obtain a refinement and new proof of the Quantum Typical Subspace Theorem 6.1.4 that does *not* use the law of large numbers and thus gives a proof of the classical typical subspace theorem without using the law of large numbers:

Theorem 7.2.6 (Geometric Typical Subspace Theorem). *Given $\rho \in \mathcal{D}(\mathcal{H})$ and $\epsilon, \delta > 0$, there exists d_0 such that for all $d \geq d_0$, there exists an orthogonal direct sum decomposition $\mathcal{H}^{\otimes d} = \mathcal{H}^d_{\epsilon-\mathrm{typ}} \oplus \mathcal{H}^d_{\delta-\mathrm{small}}$ satisfying:*

(1) $\Pr(\mathcal{H}^d_{\delta-\mathrm{small}}) < \delta$; *i.e.,* $\operatorname{trace}(\operatorname{Proj}_{\mathcal{H}^d_{\delta-\mathrm{small}}} \rho^{\otimes d}) \leq \delta$.

(2) $(1-\delta)2^{d(H(\rho)-\epsilon)} \leq \dim(\mathcal{H}^d_{\epsilon-\mathrm{typ}}) \leq 2^{d(H(\rho)+\epsilon)}$.

Moreover, one may take $\mathcal{H}^d_{\epsilon-\mathrm{typ}} = \mathrm{Rep}_{\epsilon,d,\operatorname{spec}(\rho)}$.

Proof. First note that by (7.1.7) and the observation that the number of relevant partitions is bounded above by d^n, one has

$$\begin{aligned}\dim(\mathrm{Rep}_{\epsilon,d,\mathrm{spec}(\rho)}) &< (d+1)^{\binom{n}{2}} 2^{dH(\rho)} d^n \\ &< 2^{d(H(\rho)+\frac{1}{d}n^3\log(d))}.\end{aligned}$$

Take d sufficiently large that $\epsilon > \frac{1}{d}n^3\log(d)$ to obtain the upper bound in (2). The lower bound is left to the reader.

Consider the orthogonal complement $\mathrm{Rep}^{\perp}_{\epsilon,d,\mathrm{spec}(\rho)}$ to $\mathrm{Rep}_{\epsilon,d,\mathrm{spec}(\rho)} \subset \mathcal{H}^{\otimes d}$. We must show that, given $\delta > 0$, for d sufficiently large one has

$$\text{(7.2.1)} \qquad \mathrm{trace}(\mathrm{Proj}_{\mathrm{Rep}^{\perp}_{\epsilon,d,\mathrm{spec}(\rho)}} \rho^{\otimes d}) \le \delta.$$

Let $\pi \in \mathrm{Rep}^{\perp}_{\epsilon,d,\mathrm{spec}(\rho)}$, so $\|\bar{\pi} - \mathrm{spec}(\rho)\|_1 > \epsilon$. Using Theorem 7.2.1 and then Pinsker's inequality (Lemma 6.2.5),

$$\begin{aligned}\mathrm{trace}(\mathrm{Proj}_{\pi}\rho^{\otimes d}) &\le (d+1)^{\binom{n}{2}} 2^{-dH(\bar{\pi}\|\mathrm{spec}(\rho))} \\ &\le (d+1)^{\binom{n}{2}} 2^{-d\frac{\epsilon}{2\ln(2)}}.\end{aligned}$$

Using again that the number of such π is bounded above by d^n gives

$$\mathrm{trace}(\mathrm{Proj}_{\mathrm{Rep}^{\perp}_{\epsilon,d,\mathrm{spec}(\rho)}} \rho^{\otimes d}) \le d^{n^3} 2^{-d\frac{\epsilon}{2\ln(2)}},$$

which proves that (7.2.1) holds for sufficiently large d. □

7.3. The quantum marginal problem

What are the relations between the spectrum of a density operator $\rho_{AB} \in \mathcal{D}(\mathcal{H}_A \otimes \mathcal{H}_B)$ and the spectra of its marginals $\rho_A = \mathrm{trace}_{\mathcal{H}_B}\rho_{AB}$, $\rho_B = \mathrm{trace}_{\mathcal{H}_A}\rho_{AB}$?

We may study the problem in symmetric form by introducing a purification (Definition 5.2.10) $\mathcal{H}_C \simeq \mathcal{H}_A \otimes \mathcal{H}_B$: given $|\psi\rangle \in \mathcal{H}_A \otimes \mathcal{H}_B \otimes \mathcal{H}_C$, consider

$$\begin{aligned}\rho_A &= \mathrm{trace}_{\mathcal{H}_B\otimes\mathcal{H}_C} |\psi\rangle\langle\psi|, \\ \rho_B &= \mathrm{trace}_{\mathcal{H}_A\otimes\mathcal{H}_C} |\psi\rangle\langle\psi|, \text{ and} \\ \rho_C &= \mathrm{trace}_{\mathcal{H}_A\otimes\mathcal{H}_B} |\psi\rangle\langle\psi|.\end{aligned}$$

The proof of Proposition 6.1.2(3) shows that $\mathrm{spec}(\rho_C) \cong \mathrm{spec}(\rho_{AB})$.

Problem. *Determine the relations among the spectra of ρ_A, ρ_B, ρ_C.*

Our geometric quantum typical subspace theorem (Theorem 7.2.6) indicates that the answer should be related to representation theory. We will obtain the relations by comparing the spectra of $\rho_A^{\otimes d}, \rho_B^{\otimes d}, \rho_C^{\otimes d}$ for large d and observe that they are subject to two different types of restrictions. First, $\rho_A^{\otimes d}$ projects mostly onto the submodule $\text{Rep}_{\epsilon,d,\text{spec}(\rho_A)}$ and analogously for $\rho_B^{\otimes d}, \rho_C^{\otimes d}$. Second, $\rho_{ABC} = |\psi\rangle\langle\psi|$, so $\rho_{ABC}^{\otimes d} = |\psi\rangle^{\otimes d}\langle\psi|^{\otimes d} \in \text{End}(S^d(\mathcal{H}_A \otimes \mathcal{H}_B \otimes \mathcal{H}_C))$, and the restriction of $\text{End}((\mathcal{H}_A \otimes \mathcal{H}_B \otimes \mathcal{H}_C)^{\otimes d})$ to $\text{End}(S^d(\mathcal{H}_A \otimes \mathcal{H}_B \otimes \mathcal{H}_C))$ will place constraints on the possible triples.

Recall that $S^d V = (V^{\otimes d})^{\mathfrak{S}_d}$. We have

$$\begin{aligned}
S^d(\mathcal{H}_A \otimes \mathcal{H}_B \otimes \mathcal{H}_C) &= [(\mathcal{H}_A \otimes \mathcal{H}_B \otimes \mathcal{H}_C)^{\otimes d}]^{\mathfrak{S}_d} \\
&= [\mathcal{H}_A^{\otimes d} \otimes \mathcal{H}_B^{\otimes d} \otimes \mathcal{H}_C^{\otimes d}]^{\mathfrak{S}_d} \\
&= \left[\left(\bigoplus_\mu S_\mu \mathcal{H}_A \otimes [\mu]\right) \otimes \left(\bigoplus_\nu S_\nu \mathcal{H}_B \otimes [\nu]\right) \otimes \left(\bigoplus_\pi S_\pi \mathcal{H}_C \otimes [\pi]\right)\right]^{\mathfrak{S}_d} \\
&= \bigoplus_{\mu,\nu,\pi} (S_\mu \mathcal{H}_A \otimes S_\nu \mathcal{H}_B \otimes S_\pi \mathcal{H}_C) \otimes ([\mu] \otimes [\nu] \otimes [\pi])^{\mathfrak{S}_d}.
\end{aligned}$$

On one hand, the modules $S_\mu \mathcal{H}_A \otimes [\mu]$ for which $\dim(\text{Proj}_\mu(\rho_A^{\otimes d}))$ is large must have $\bar{\mu}$ close to $\text{spec}(\rho_A)$, and analogously for ρ_B, ρ_C. On the other hand, the triples μ, ν, π appearing must be such that $[\mu] \otimes [\nu] \otimes [\pi]$ contains a trivial $\mathfrak{S}_d$-module. This tension will be the key to the solution of the problem.

Definition 7.3.1. Define the *Kronecker coefficients*

$$k_{\mu\nu\pi} := \dim([\mu] \otimes [\nu] \otimes [\pi])^{\mathfrak{S}_d}.$$

There is an extensive literature on Kronecker coefficients, but they are not well understood. See, e.g., [**Man15a**] and the references therein. It is not even well understood when they are zero.

While $\bar{\pi}, \bar{\mu}, \bar{\nu}$ must be rational vectors, triples of spectra are not subject to such restrictions. This indicates that the results about the marginals will have to involve sequences.

7.3.1. Statement of results.

For the rest of §7.3, fix $n = \dim \mathcal{H}_A$, $m = \dim \mathcal{H}_B$, $k = \dim \mathcal{H}_C$.

Theorem 7.3.2 ([**CM06**]). *Let ρ_{AB} be a density operator on $\mathcal{H}_A \otimes \mathcal{H}_B$. Then there exists a sequence (π_d, μ_d, ν_d) of triples of partitions such that $k_{\pi_d,\mu_d,\nu_d} \neq 0$ for all d and*

$$\begin{aligned}
\lim_{d\to\infty} \overline{\pi}_d &= \text{spec}(\rho_{AB}), \\
\lim_{d\to\infty} \overline{\mu}_d &= \text{spec}(\rho_A), \\
\lim_{d\to\infty} \overline{\nu}_d &= \text{spec}(\rho_B).
\end{aligned}$$

Theorem 7.3.3 ([**Kly04**]). *Let ρ_{AB} be a density operator on $\mathcal{H}_A \otimes \mathcal{H}_B$ such that* $\mathrm{spec}(\rho_{AB})$, $\mathrm{spec}(\rho_A)$, *and* $\mathrm{spec}(\rho_B)$ *are all rational vectors. Then there exists an $M > 0$ such that $k_{M\,\mathrm{spec}(\rho_A), M\,\mathrm{spec}(\rho_B), M\,\mathrm{spec}(\rho_{AB})} \neq 0$.*

Theorem 7.3.4 ([**Kly04**]). *Let π, μ, ν be partitions of d with $k_{\pi\mu\nu} \neq 0$ and satisfying $\ell(\pi) \leq mn$, $\ell(\mu) \leq m$, and $\ell(\nu) \leq n$. Then there exists a density operator ρ_{AB} on $\mathbb{C}^n \otimes \mathbb{C}^m = \mathcal{H}_A \otimes \mathcal{H}_B$ with* $\mathrm{spec}(\rho_{AB}) = \bar{\pi}$, $\mathrm{spec}(\rho_A) = \bar{\mu}$, *and* $\mathrm{spec}(\rho_B) = \bar{\nu}$.

Klyachko's proofs are via co-adjoint orbits and vector bundles on flag varieties, while the proof of Christandl-Mitchison is information-theoretic. We give information-theoretic proofs of all three theorems following [**CHM07**].

Remark 7.3.5. As remarked above, the nonvanishing of Kronecker coefficients is not well understood, so the results do not completely solve the problem. However, much qualitative information is known, and they are understood in small cases; e.g., $m = n = k = 3$ [**Kly04**]. See [**Res10, Res11**] for a geometric perspective on nonvanishing.

7.3.2. Proof of Theorem 7.3.2.

Lemma 7.3.6. *For all projection operators $P \in \mathrm{End}(\mathcal{H}_A)$, $Q \in \mathrm{End}(\mathcal{H}_B)$, $R \in \mathrm{End}(\mathcal{H}_C)$, and density operators ρ_{ABC} on $\mathcal{H}_A \otimes \mathcal{H}_B \otimes \mathcal{H}_C$,*

$$\mathrm{trace}((P \otimes Q)\rho_{AB}) \geq \mathrm{trace}(P\rho_A) + \mathrm{trace}(Q\rho_B) - 1, \tag{7.3.1}$$

and

$$\mathrm{trace}((P \otimes Q \otimes R)\rho_{ABC}) \geq \mathrm{trace}(P\rho_A) + \mathrm{trace}(Q\rho_B) + \mathrm{trace}(R\rho_C) - 2. \tag{7.3.2}$$

Proof. Consider

$$\begin{aligned}
0 &\leq \mathrm{trace}\big[\big((\mathrm{Id}_{\mathcal{H}_A} - P) \otimes (\mathrm{Id}_{\mathcal{H}_B} - Q)\big)\,\rho_{AB}\big] \\
&= \mathrm{trace}[\rho_{AB} - P \otimes \mathrm{Id}_{\mathcal{H}_B}(\rho_{AB}) - \mathrm{Id}_{\mathcal{H}_A} \otimes Q(\rho_{AB}) + (P \otimes Q)\rho_{AB}] \\
&= 1 - \mathrm{trace}(P\rho_A) - \mathrm{trace}(Q\rho_B) + \mathrm{trace}((P \otimes Q)\rho_{AB}),
\end{aligned}$$

proving the first assertion. To prove the second assertion, apply the first to $\mathrm{trace}(((P \otimes Q) \otimes R)\rho_{ABC})$. □

Proof of Theorem 7.3.2. It will be convenient to view the theorem symmetrically by taking a purification $|\psi\rangle \in \mathcal{H}_A \otimes \mathcal{H}_B \otimes \mathcal{H}_C$, so

$$\rho_{AB} \cong \rho_C = \mathrm{trace}_{\mathcal{H}_A \otimes \mathcal{H}_B}(|\psi\rangle\langle\psi|).$$

Since we are working symmetrically, we might as well allow any dimension for $\mathcal{H}_C$ and prove a slightly stronger result. By Theorem 7.2.6, for all $\epsilon, \delta > 0$, there exists d_0 such that for all $d \geq d_0$

$$\begin{aligned} &\operatorname{trace}(\operatorname{Proj}_{\operatorname{Rep}_{\epsilon,d,\operatorname{spec}(\rho_A)}} \rho_A^{\otimes d}) \geq 1-\delta, \\ &\operatorname{trace}(\operatorname{Proj}_{\operatorname{Rep}_{\epsilon,d,\operatorname{spec}(\rho_B)}} \rho_B^{\otimes d}) \geq 1-\delta, \qquad (7.3.3) \\ &\operatorname{trace}(\operatorname{Proj}_{\operatorname{Rep}_{\epsilon,d,\operatorname{spec}(\rho_C)}} \rho_C^{\otimes d}) \geq 1-\delta. \end{aligned}$$

Apply Lemma 7.3.6 to obtain

$$\begin{aligned} &\operatorname{trace}\Big(\operatorname{Proj}_{\operatorname{Rep}_{\epsilon,d,\operatorname{spec}(\rho_A)}} \otimes \operatorname{Proj}_{\operatorname{Rep}_{\epsilon,d,\operatorname{spec}(\rho_B)}} \otimes \operatorname{Proj}_{\operatorname{Rep}_{\epsilon,d,\operatorname{spec}(\rho_C)}} (|\psi\rangle\langle\psi|)^{\otimes d}\Big) \\ &\geq 3(1-\delta)-2 = 1-3\delta. \end{aligned}$$

Assuming $\delta < \frac{1}{3}$, since each of $\operatorname{Proj}_{\operatorname{Rep}_{\epsilon,d,\operatorname{spec}(\rho_A)}}$, $\operatorname{Proj}_{\operatorname{Rep}_{\epsilon,d,\operatorname{spec}(\rho_B)}}$, $\operatorname{Proj}_{\operatorname{Rep}_{\epsilon,d,\operatorname{spec}(\rho_C)}}$ is a sum of projection operators, there must be one triple (μ,ν,π), with $\bar{\mu} \in B_\epsilon(\operatorname{spec}(\rho_A))$, $\bar{\nu} \in B_\epsilon(\operatorname{spec}(\rho_B))$, $\bar{\pi} \in B_\epsilon(\operatorname{spec}(\rho_C))$, with $k_{\pi\mu\nu} \neq 0$ for which the projection is nonzero. Now just take a sequence of such as $\epsilon \to 0$. □

7.3.3. Consequences of Theorem 7.3.2. By its definition, $k_{\pi\mu\nu}$ is symmetric in its arguments. Representations of the symmetric group are self-dual: as $\mathfrak{S}_d$-modules, $[\pi]^* \cong [\pi]$, but there is no canonical $\mathfrak{S}_d$-isomorphism between the two spaces.

Thus

$$\begin{aligned} ([\pi]\otimes[\mu]\otimes[\mu])^{\mathfrak{S}_d} &\cong ([\pi]^*\otimes[\mu]\otimes[\nu])^{\mathfrak{S}_d} \\ &= \operatorname{Hom}_{\mathfrak{S}_d}([\pi],[\mu]\otimes[\nu]); \end{aligned}$$

i.e.,

$$k_{\pi\mu\nu} = \operatorname{mult}([\pi],[\mu]\otimes[\nu]). \qquad (7.3.4)$$

Exercise 7.3.7. Fix $M \in \mathbb{N}$. Let k_d denote the largest Kronecker coefficient $k_{\pi\mu\nu}$ when $|\pi| = |\mu| = |\nu| = d$ and $\ell(\pi), \ell(\mu), \ell(\nu)$ are all at most M. Show that the growth of k_d is bounded by a polynomial in d. ⊚

Theorem 7.3.2 gives a new proof of Proposition 6.2.13 on the subadditivity of von Neumann entropy:

Corollary 7.3.8 (Subadditivity of von Neumann entropy).

$$H(\rho_{AB}) \leq H(\rho_A) + H(\rho_B).$$

Proof. By (7.3.4), $[\mu]\otimes[\nu] = \bigoplus_\pi [\pi]^{\oplus k_{\pi\mu\nu}}$. Thus if $k_{\pi\mu\nu} \neq 0$, then $\dim([\mu]\otimes[\nu]) \geq \dim[\pi]$. Now take a sequence $(\pi(j),\mu(j),\nu(j))$ of partitions of $d(j)$ with $d(j) \to \infty$, with $k_{\pi(j),\mu(j),\nu(j)} \neq 0$ and $(\bar{\pi}(j),\bar{\mu}(j),\bar{\nu}(j))$ converging to $(\operatorname{spec}(\rho_{AB}), \operatorname{spec}(\rho_A), \operatorname{spec}(\rho_B))$. In particular, for each j,

$$\dim[\pi(j)] \leq \dim[\mu(j)]\dim[\nu(j)]. \qquad (7.3.5)$$

By (7.1.6), $\frac{1}{d_j}\log(\dim[\pi(j)])$ limits to $H(\rho_{AB})$ and similarly for the other two limits. □

Compare this proof with the previous proof of the subadditivity, which relied on the nonnegativity of the relative entropy, which in turn relied on convexity. Here we obtain the result by studying the growth of dimensions of modules of the permutation group.

Similarly, we recover the triangle inequality Proposition 6.2.15 (although here the proof is essentially the same as before):

Corollary 7.3.9 (Triangle inequality for von Neumann entropy). *Let $\rho_{AB} \in \mathcal{D}(\mathcal{H}_A \otimes \mathcal{H}_B)$. Then*

$$H(\rho_{AB}) \geq |H(\rho_A) - H(\rho_B)|.$$

Proof. Take a purification so $H(\rho_{AB}) = H(\rho_C)$. By Corollary 7.3.8, $H(\rho_A) \leq H(\rho_B) + H(\rho_C)$, i.e., $H(\rho_C) \geq H(\rho_A) - H(\rho_B)$, and similarly exchanging the roles of A and B. □

7.3.4. Weights, Cartan products, and a result on Kronecker coefficients.

We need a little more representation theory before giving the proof of Theorem 7.3.3.

Note that for all I, π, the projection $\mathrm{Proj}_\pi(e_I)$ is either a weight vector of weight $\mathrm{wt}(e_I)$ or zero.

Definition 7.3.10. A weight vector $v \in (\mathbb{C}^n)^{\otimes d}$ is a *highest weight vector* if $g \cdot v = v$ for all $g \in N_n$, where $N_n \subset \mathrm{GL}_n$ is the subgroup of matrices with 1's on the diagonal and 0's below the diagonal.

Extend the notion of weight vectors and highest weight vectors to $G = \mathrm{GL}_m \times \mathrm{GL}_n \times \mathrm{GL}_k$ in the natural way, so the weight of $e_I \otimes f_J \otimes h_K$ is a triple of weights and a weight vector is a highest weight vector if it is invariant under $N_{m,n,k} := N_m \times N_n \times N_k$.

See [**Lan12**, §6.8] for a discussion of highest weight vectors in this context.

Fact. If V is an irreducible G-module, where G is GL_n, $\mathbf{U}(n)$, or products of such, then there exists a unique up to scale highest vector in V. (See, e.g., [**Kna02**, V.2].)

Exercise 7.3.11. If $v \in S_\pi\mathbb{C}^n$ is a highest weight vector, then $\mathrm{wt}(v) = \pi$. In particular, irreducible GL_n-modules are indexed by their highest weights.

Exercise 7.3.12. If $v \in V^{\otimes d}$ and $w \in V^{\otimes \delta}$ are highest weight vectors, then $v \otimes w \in V^{\otimes d+\delta}$ is a highest weight vector.

Let

$$\mathrm{Kron}_{m,n,k} := \{(\mu, \nu, \pi) \mid k_{\mu\nu\pi} \neq 0 \text{ and } \ell(\mu) \leq m,\ \ell(\nu) \leq n,\ \ell(\pi) \leq k\},$$

the set of all triples of partitions (μ, ν, π) such that $S_\mu\mathbb{C}^m \otimes S_\nu\mathbb{C}^n \otimes S_\pi\mathbb{C}^k$ appears in $\mathrm{Sym}(\mathbb{C}^m \otimes \mathbb{C}^n \otimes \mathbb{C}^k) := \bigoplus_d S^d(\mathbb{C}^m \otimes \mathbb{C}^n \otimes \mathbb{C}^k)$.

The following proposition was classically known in representation theory (the essential step in the proof dates back to E. Cartan) but unknown in the quantum information theory community until 2006:

Proposition 7.3.13. $\mathrm{Kron}_{m,n,k}$ *is a semi-group. In other words, if* $k_{\pi\mu\nu} \neq 0$ *and* $k_{\pi'\mu'\nu'} \neq 0$, *then* $k_{\pi+\pi',\mu+\mu',\nu+\nu'} \neq 0$. *In particular, if* $k_{\pi\mu\nu} \neq 0$, *then* $k_{M\pi,M\mu,M\nu} \neq 0$ *for all* $M \in \mathbb{N}$.

Proof. Let $X_{\pi,\mu,\nu} \in S^d(\mathbb{C}^k \otimes \mathbb{C}^m \otimes \mathbb{C}^n)$ be a highest weight vector for $\mathrm{GL}_k \times \mathrm{GL}_m \times \mathrm{GL}_n$ of weight (π, μ, ν), and let $X'_{\pi',\mu',\nu'} \in S^{d'}(\mathbb{C}^k \otimes \mathbb{C}^m \otimes \mathbb{C}^n)$ be a highest weight vector of weight (π', μ', ν'), both of which exist by the nonvanishing of the Kronecker coefficients in the hypothesis. The multiplication of polynomials $S^dW \times S^\delta W \to S^{d+\delta}W$ commutes with the action of $\mathrm{GL}(W)$. Thus $X_{\pi,\mu,\nu}X'_{\pi',\mu',\nu'} \in S^{d+d'}(\mathbb{C}^k \otimes \mathbb{C}^m \otimes \mathbb{C}^n)$ is nonzero and a highest weight vector of weight $(\pi + \pi', \mu + \mu', \nu + \nu')$. (It is a highest weight vector because already $X_{\pi,\mu,\nu} \otimes X'_{\pi',\mu',\nu'}$ is one.) □

Proposition 7.3.14 (**[CHM07]**). *If* $k_{\pi\mu\nu} \neq 0$, *then* $H(\overline{\pi}) \leq H(\overline{\mu}) + H(\overline{\nu})$.

Proof. Say $|\pi| = |\mu| = |\nu| = d$. By the semi-group property $k_{N\pi,N\mu,N\nu} \neq 0$; thus $[N\pi] \subset [N\mu] \otimes [N\nu]$ and $\dim[N\pi] \leq \dim([N\mu] \otimes [N\nu])$. Taking logs and using Lemma 7.1.24 we obtain

$$NdH(\bar{\pi}) - o(N) \leq NdH(\bar{\mu}) + NdH(\bar{\nu}).$$

Now divide by dN and let $N \to \infty$ to conclude. □

The following theorem dates back to 1967 but was unknown in the quantum information community until 2006. In fact it is conjectured in **[Kly04]** and a proof was given in **[CHM07]**:

Theorem 7.3.15 (**[Had67]**; see also **[HD73]**). $\mathrm{Kron}_{m,n,k}$ *is finitely generated.*

A proof, assuming basic commutative algebra and representation theory, is given in §7.5. The essential point is that $\mathrm{Kron}_{m,n,k}$ is the subring of the ring of polynomials on $\mathbb{C}^m \otimes \mathbb{C}^n \otimes \mathbb{C}^k$ that are invariant under the group $N_{m,n,k}$, and such rings are finitely generated by general results.

Let

$$\mathrm{KRON}_{m,n,k} := \{(\bar{\mu}, \bar{\nu}, \bar{\pi}) \mid (\mu, \nu, \pi) \in \mathrm{Kron}_{m,n,k}\} \tag{7.3.6}$$

denote the set of normalized Kronecker coefficients. By definition it consists of rational points.

Theorem 7.3.16 ([**Bri87**, §2, Prop.]). $\mathrm{KRON}_{m,n,k}$ *is a rational convex polytope, i.e., the rational points in the convex hull of a finite number of points in* $\mathbb{Q}^m \times \mathbb{Q}^n \times \mathbb{Q}^k$.

Proof. Let $P_1, \dots, P_r$ be generators of the semi-group of highest weight vectors of polynomials on $\mathbb{C}^m \otimes \mathbb{C}^n \otimes \mathbb{C}^k$, i.e., $\mathrm{Sym}(\mathbb{C}^m \otimes \mathbb{C}^n \otimes \mathbb{C}^k)^{N_{m,n,k}}$, which is also denoted $\mathbb{C}[\mathbb{C}^m \otimes \mathbb{C}^n \otimes \mathbb{C}^k]^{N_{m,n,k}}$[1]. Each P_i is a highest weight vector of weight (π_i, μ_i, ν_i) for some triple of partitions of some d_i. As a semi-group, $\mathbb{C}[\mathbb{C}^m \otimes \mathbb{C}^n \otimes \mathbb{C}^k]^{N_{m,n,k}}$ coincides with $\mathrm{Kron}_{m,n,k}$.

Claim: $\mathrm{KRON}_{m,n,k}$ is the $\mathbb{Q}$-convex hull of the $(\frac{\pi_i}{d_i}, \frac{\mu_i}{d_i}, \frac{\nu_i}{d_i})$, $1 \le i \le r$.

To prove the claim, let $\beta \in \mathrm{KRON}_{m,n,k}$, so there exists $d \in \mathbb{N}$ such that $d\beta = (\pi, \mu, \nu)$ for some partitions π, μ, ν and $k_{\pi\mu\nu} \neq 0$. Thus there exists $(\delta_1, \dots, \delta_r)$ such that $P_1^{\otimes \delta_1} \otimes \cdots \otimes P_r^{\otimes \delta_r} \in (\mathbb{C}^m \otimes \mathbb{C}^n \otimes \mathbb{C}^k)^{\otimes d}$ has weight $d\beta$. I.e., $d\beta = (\delta_1\pi_1 + \cdots + \delta_r\pi_r, \delta_1\mu_1 + \cdots + \delta_r\mu_r, \delta_1\nu_1 + \cdots + \delta_r\nu_r)$, so

$$\beta = \frac{d_1\delta_1}{d}\left(\frac{\pi_1}{d_1}, \frac{\mu_1}{d_1}, \frac{\nu_1}{d_1}\right) + \cdots + \frac{d_r\delta_r}{d}\left(\frac{\pi_r}{d_r}, \frac{\mu_r}{d_r}, \frac{\nu_r}{d_r}\right).$$

Finally note that $d_1\delta_1 + \cdots + d_r\delta_r = d$, showing the linear combination is indeed convex. □

7.3.5. Proofs of Klyachko's theorems. Let $\mathrm{SPEC}_{m,n,k}$ denote the set of possible triples of spectra of marginals of rank one density operators on $\mathbb{C}^m \otimes \mathbb{C}^n \otimes \mathbb{C}^k$.

Proof of Theorem 7.3.3. Let $(r_A, r_B, r_C) \in \mathrm{SPEC}_{m,n,k}$ with each vector rational. Theorem 7.3.2 implies there exists a sequence $(\mu(j), \nu(j), \pi(j)) \in \mathrm{Kron}_{m,n,k}$ with $(\bar{\mu}(j), \bar{\nu}(j), \bar{\pi}(j)) \in \mathrm{KRON}_{m,n,k}$ converging to (r_A, r_B, r_C). The rationality of (r_A, r_B, r_C) and Theorem 7.3.16 imply $(r_A, r_B, r_C) \in \mathrm{KRON}_{m,n,k}$. Thus there exists an M such that $(Mr_A, Mr_B, Mr_C) \in \mathrm{Kron}_{m,n,k}$. □

Exercise 7.3.17. As mentioned above, Theorem 7.3.3 was originally proved using different methods. Use Theorem 7.3.3 to give a second proof of Theorem 7.3.2.

To prove Theorem 7.3.4 we will need the following lemma:

Lemma 7.3.18. *Let* $(\mu, \nu, \pi) \in \mathrm{Kron}_{m,n,k}$ *with* $|\mu| = d$. *Then there exists a unit vector* $|\psi\rangle \in \mathcal{H}_A \otimes \mathcal{H}_B \otimes \mathcal{H}_C$ *such that the marginals of the corresponding density operator* $\rho_{ABC} = |\psi\rangle\langle\psi|$ *satisfy*

$$\max\{\|\mathrm{spec}(\rho_A) - \bar{\mu}\|_1, \|\mathrm{spec}(\rho_B) - \bar{\nu}\|_1, \|\mathrm{spec}(\rho_C) - \bar{\pi}\|_1\}$$
$$\le \mathrm{poly}(m, n, k)\sqrt{\frac{\log(d)}{d}}.$$

[1] There is a slight difference as the second notation *a priori* does not involve the grading.

Here is an outline of the proof: we first fix a unit element $|v\rangle \in S_\mu\mathbb{C}^m\otimes S_\nu\mathbb{C}^n\otimes S_\pi\mathbb{C}^k$ and show there is some $|\psi\rangle \in \mathbb{C}^m\otimes\mathbb{C}^n\otimes\mathbb{C}^k$ with $|\psi^{\otimes d}\rangle$ close to $|v\rangle$. We use $\mathrm{trace}(\mathrm{Proj}_{\mathrm{span}\,|v\rangle\langle v|}(|\psi^d\rangle\langle\psi^d|))$ to lower bound the trace of $\mathrm{Proj}_\mu\rho_A^{\otimes d}$. We upper bound this trace using the spectrum estimation theorem in terms of the relative entropy and conclude by applying Pinsker's inequality. Here are the details:

Proof. Let

$$|v\rangle \in S_\mu\mathbb{C}^m\otimes S_\nu\mathbb{C}^n\otimes S_\pi\mathbb{C}^k \subset S^d(\mathbb{C}^m\otimes\mathbb{C}^n\otimes\mathbb{C}^k)$$

have length one. We want to find $|\psi\rangle \in \mathbb{C}^m\otimes\mathbb{C}^n\otimes\mathbb{C}^k =: W$ such that $|\psi\rangle^{\otimes d}$ is close to $|v\rangle$. The variety of d-th powers of linear forms, called the *Veronese variety* and denoted $\hat{v}_d(\mathbb{P}W)$, spans S^dW. One may take an orthonormal basis of S^dW from elements of $\hat{v}_d(\mathbb{P}W)$. Taking one such basis, at worst $|v\rangle$ is a sum of $\dim S^dW$ vectors in $\hat{v}_d(\mathbb{P}W)$, all equidistant from $|v\rangle$, so taking $|\psi\rangle^{\otimes d}$, the closest such basis vector to $|v\rangle$, we certainly may attain

$$|\langle\psi^{\otimes d}|v\rangle| \geq \frac{1}{\sqrt{\dim S^dW}} > \frac{1}{d^{\frac{mnk}{2}}}.$$

Now $|v\rangle$ is just one vector in $S_\mu\mathbb{C}^m\otimes S_\nu\mathbb{C}^n\otimes S_\pi\mathbb{C}^k$, so

$$\begin{aligned}\mathrm{trace}(\mathrm{Proj}_\mu\otimes\mathrm{Proj}_\nu\otimes\mathrm{Proj}_\pi(|\psi\rangle\langle\psi|)^{\otimes d}) &\geq \mathrm{trace}(|v\rangle\langle v|(|\psi\rangle\langle\psi|)^{\otimes d})\\ &= |\langle\psi^{\otimes d}|v\rangle|^2\\ &> \frac{1}{d^{mnk}}.\end{aligned}$$

In particular,

$$\mathrm{trace}(\mathrm{Proj}_\mu\rho_A^{\otimes d}) = \mathrm{trace}\big((\mathrm{Proj}_\mu\otimes\mathrm{Id}_{\mathbb{C}^n}\otimes\mathrm{Id}_{\mathbb{C}^k})\rho_{ABC}^{\otimes d}\big) > \frac{1}{d^{mnk}}.$$

On the other hand, the Spectrum Estimation Theorem 7.2.1 says

$$\mathrm{trace}(\mathrm{Proj}_\mu\rho_A^{\otimes d}) \leq (d+1)^{\binom{m}{2}}2^{-dH(\bar{\mu}\|\,\mathrm{spec}(\rho_A))}.$$

Combining the two inequalities and taking logs gives

$$H(\bar{\mu}\|\,\mathrm{spec}(\rho_A)) \leq \frac{1}{d}\left[\binom{m}{2}\log(d+1) + mnk\log d\right].$$

Finally, Pinsker's inequality says

$$H(\bar{\mu}\|\operatorname{spec}(\rho_A)) \geq \frac{1}{2\ln(2)}\|\bar{\mu}-\operatorname{spec}(\rho_A)\|_1^2,$$

proving the polynomial bound for $\|\bar{\mu}-\operatorname{spec}(\rho_A)\|_1^2$. Applying the same argument to the B, C factors we conclude. □

Proof of Theorem 7.3.4. For all N, $(N\mu, N\nu, N\pi) \in \mathrm{Kron}_{m,n,k}$; thus by Lemma 7.3.18 we may obtain a sequence $\{|\psi_N\rangle\}$ whose associated density operators have marginals converging to $(\bar{\mu}, \bar{\nu}, \bar{\pi})$. Since the set of unit vectors is compact we obtain a limiting density operator with the desired marginals. □

Together the theorems imply:

Theorem 7.3.19 ([**CHM07**]).

$$\mathrm{SPEC}_{m,n,k} = \overline{\mathrm{KRON}_{m,n,k}}.$$

7.4. Tripartite states

Let ρ_{ABC} be a density operator on $\mathcal{H}_A \otimes \mathcal{H}_B \otimes \mathcal{H}_C$ (equivalently, take some unit vector $|\psi\rangle \in \mathcal{H}_A \otimes \mathcal{H}_B \otimes \mathcal{H}_C \otimes \mathcal{H}_D$). What are the relations between the marginal density operators? In particular, we are interested in entropy relations, i.e., inequalities among

$$H(\rho_{ABC}), H(\rho_{AB}), \ldots, H(\rho_A), H(\rho_B), H(\rho_C).$$

Above we treated the case where ρ_{ABC} is pure.

Let $a = \dim \mathcal{H}_A$, $b = \dim \mathcal{H}_B$, $c = \dim \mathcal{H}_C$. Introduce the notation $\alpha \vdash_a d$ to mean α is a partition of d with at most a parts. Consider the $\mathfrak{S}_d \times \mathbf{U}(\mathcal{H}_A) \times \mathbf{U}(\mathcal{H}_B) \times \mathbf{U}(\mathcal{H}_C)$ decomposition

$$(\mathcal{H}_A \otimes \mathcal{H}_B \otimes \mathcal{H}_C)^{\otimes d} = \bigoplus_{\alpha \vdash_a d, \beta \vdash_b d, \gamma \vdash_c d} ([\alpha] \otimes [\beta] \otimes [\gamma]) \otimes (S_\alpha \mathcal{H}_A \otimes S_\beta \mathcal{H}_B \otimes S_\gamma \mathcal{H}_C).$$

As with the previous case, if we take a purification to get $\rho_{ABCD} = |\psi\rangle\langle\psi|$ and $\operatorname{spec}(\rho_{ABC}) = \operatorname{spec}(\rho_D)$, etc., there are restrictions on the spectra: (i) the geometric typical subspace theorem says the isotypic components corresponding to partitions near the spectra absorb most of the probability and (ii) if a 4-tuple of partitions $\alpha, \beta, \gamma, \delta$ appears, then $([\alpha] \otimes [\beta] \otimes [\gamma] \otimes [\delta])^{\mathfrak{S}_d}$ must be nonzero. Here there are also intermediate operators ρ_{AB}, and there will be additional restrictions taking those into account, which we now describe.

7.4.1. Tripartite typical subspace theorem. Consider the decomposition of $[\alpha]\otimes[\beta]\otimes[\gamma]$ under the permutation group $\mathfrak{S}_d$. Two ways to arrive at it in two steps are the following left or right downward pointing arrows:
(7.4.1)

$$\begin{array}{ccc} & [\alpha]\otimes[\beta]\otimes[\gamma] & \\ \swarrow & & \searrow \\ [\alpha]\otimes([\beta]\otimes[\gamma]) & & ([\alpha]\otimes[\beta])\otimes[\gamma] \\ \| & & \| \\ [\alpha]\otimes\left(\oplus_\nu[\nu]\otimes K_\nu^{\beta\gamma}\right) & & \left(\oplus_\mu[\mu]\otimes K_\mu^{\alpha\beta}\right)\otimes[\gamma] \\ \searrow & & \swarrow \\ & \bigoplus_{[\lambda]}[\lambda]\otimes K_\lambda^{\alpha\beta\gamma} & \end{array}$$

Here $K_\lambda^{\beta\gamma} = \mathrm{Hom}_{\mathfrak{S}_d}([\lambda],[\beta]\otimes[\gamma]) = \mathbb{C}^{k_{\lambda\beta\gamma}}$ is a space recording the multiplicity of $[\lambda]$ in $[\beta]\otimes[\gamma]$ and similarly for $K_\mu^{\alpha\beta}$, and $K_\lambda^{\alpha\beta\gamma} = \mathrm{Hom}_{\mathfrak{S}_d}([\lambda],[\alpha]\otimes[\beta]\otimes[\gamma])$. Focus on the isotypic component of some fixed $[\lambda]$.

Similarly to the Choi-Jamiołkowski isomorphism, for each partition α, choose an $\mathfrak{S}_d$-isomorphism $\tau : [\alpha] \to [\alpha]^*$. We may use these choices to reverse the arrows. Let

$$\mathrm{Iso}_\lambda^{\alpha\beta\gamma} : [\lambda]\otimes K_\lambda^{\alpha\beta\gamma} \to [\lambda]\otimes K_\lambda^{\alpha\beta\gamma}$$

denote the isomorphism obtained by traveling the circuit counterclockwise, so it depends on our choices of $\mathfrak{S}_d$-isomorphisms, although the results about dimensions will not. An analog of the typical subspace theorem will identify the contribution of any fixed pair of partitions μ, ν to the circuit. That is, fix some $[\mu]$ and some $[\nu]$. We obtain restricted maps

$$\begin{array}{ccc} & [\alpha]\otimes[\beta]\otimes[\gamma] & \\ \swarrow & & \searrow \\ [\alpha]\otimes[\nu]\otimes K_\nu^{\beta\gamma} & & [\mu]\otimes K_\mu^{\alpha\beta}\otimes[\gamma] \\ \searrow & & \swarrow \\ & [\lambda]\otimes K_\lambda^{\alpha\beta\gamma} & \end{array} \tag{7.4.2}$$

and, reversing arrows on the right and traveling counterclockwise, a restricted map

$$\mathrm{Id}_\lambda \otimes \begin{bmatrix} \alpha\beta\gamma \\ \mu\nu \\ \lambda \end{bmatrix} : [\lambda]\otimes K_\lambda^{\alpha\beta\gamma} \to [\lambda]\otimes K_\lambda^{\alpha\beta\gamma},$$

so that

$$\mathrm{Iso}_\lambda^{\alpha\beta\gamma} = \bigoplus_{\mu,\nu} \left(\mathrm{Id}_\lambda \otimes \begin{bmatrix} \alpha\beta\gamma \\ \mu\nu \\ \lambda \end{bmatrix} \right).$$

Definition 7.4.1. The linear maps

$$\begin{bmatrix} \alpha\beta\gamma \\ \mu\nu \\ \lambda \end{bmatrix} : K_\lambda^{\alpha\beta\gamma} \to K_\lambda^{\alpha\beta\gamma}$$

are called *recoupling coefficients*.

The name comes from the physics literature where they work in bases, and the coefficients refer to the entries of the matrices representing the linear maps.

The next theorem shows that for any density operator ρ_{ABC} on $\mathcal{H}_A \otimes \mathcal{H}_B \otimes \mathcal{H}_C$ and for large d, the set of normalized 6-tuples of partitions $(\bar{\alpha}, \bar{\beta}, \bar{\gamma}, \bar{\mu}, \bar{\nu}, \bar{\lambda})$ that are δ-close to

$$(\mathrm{spec}(\rho_A), \mathrm{spec}(\rho_B), \mathrm{spec}(\rho_C), \mathrm{spec}(\rho_{AB}), \mathrm{spec}(\rho_{BC}), \mathrm{spec}(\rho_{ABC}))$$

contains a fraction of at least $1 - \epsilon$ of the isomorphism $\mathrm{Iso}_\lambda^{\alpha\beta\gamma}$, where $\epsilon \sim 2^{-d}$. Since there are only $\mathrm{poly}(d)$ such close partitions, on average one captures a fraction on the order of $\frac{1}{\mathrm{poly}(d)}$ of the isomorphism, and in particular, there is at least one that captures $\frac{1}{\mathrm{poly}(d)}$ of the isomorphism. Conversely, for any sequence of 6-tuples not in some ϵ-ball of some 6-tuple of marginal spectra, the total fraction of the isomorphism it captures is exponentially small.

Recall the trace norm $||A||_1 := \mathrm{trace}(\sqrt{A^\dagger A})$ from Definition 5.3.1.

Theorem 7.4.2 (**[CcW18]**). *Let ρ_{ABC} be a density operator on $\mathcal{H}_A \otimes \mathcal{H}_B \otimes \mathcal{H}_C$. Then there exists a sequence of partitions $(\alpha_d, \beta_d, \gamma_d, \mu_d, \nu_d, \lambda_d)$ such that the normalized sequence $(\bar{\alpha}_d, \bar{\beta}_d, \bar{\gamma}_d, \bar{\mu}_d, \bar{\nu}_d, \bar{\lambda}_d)$ converges to*

$$(\mathrm{spec}(\rho_A), \mathrm{spec}(\rho_B), \mathrm{spec}(\rho_C), \mathrm{spec}(\rho_{AB}), \mathrm{spec}(\rho_{BC}), \mathrm{spec}(\rho_{ABC})) \tag{7.4.3}$$

and

$$\left\| \begin{bmatrix} \alpha_d\beta_d\gamma_d \\ \mu_d\nu_d \\ \lambda_d \end{bmatrix} \right\|_1^2 \geq \frac{1}{\mathrm{poly}(d)}.$$

Conversely, if a 6-tuple of probability distributions does not correspond to the marginals of a density operator as in (7.4.3), *then any sequence $(\bar{\alpha}_d, \dots, \bar{\lambda}_d)$ of normalized partitions converging to it satisfies*

$$\left\| \begin{bmatrix} \alpha_d\beta_d\gamma_d \\ \mu_d\nu_d \\ \lambda_d \end{bmatrix} \right\|_1^2 < 2^{-\Omega(d)}.$$

For the proof, see [**CcW18**]. It follows similar lines to the proofs in the quantum marginal problem, but the details are more complicated.

7.4.2. Symmetries of recoupling coefficients. Just as the Kronecker coefficients have an $\mathfrak{S}_3$ symmetry, the recoupling coefficients have symmetry. To see the symmetry, we first need to have numbers rather than linear maps, as the maps act on different spaces of different dimensions. For numbers we utilize the squared trace norm of the recoupling coefficients. Pictorially, this is the trace of the map by traveling around the circuit counterclockwise, then clockwise. So we look in $([\alpha]\otimes[\beta]\otimes[\gamma]\otimes[\mu]\otimes[\nu]\otimes[\lambda])^{\otimes 2}$, and the total map corresponds to a sequence of projection operators. Let

$$P^{\alpha\beta\gamma} : [\alpha]\otimes[\beta]\otimes[\gamma] \to ([\alpha]\otimes[\beta]\otimes[\gamma])^{\mathfrak{S}_d}$$

and

$$P^{\alpha\alpha} : [\alpha]\otimes[\alpha] \to ([\alpha]\otimes[\alpha])^{\mathfrak{S}_d}$$

denote the projection operators.

Define

$$\begin{aligned} P^{\alpha\beta\gamma\lambda\mu\nu} &:= (P^{\alpha\alpha}\otimes P^{\beta\beta}\otimes P^{\gamma\gamma}\otimes P^{\lambda\lambda}\otimes P^{\mu\mu}\otimes P^{\nu\nu})(P^{\mu\gamma\lambda}\otimes P^{\alpha\beta\mu}\otimes P^{\alpha\nu\lambda}\otimes P^{\beta\gamma\nu}) \\ &\in \operatorname{End}(([\alpha]\otimes[\beta]\otimes[\gamma]\otimes[\mu]\otimes[\nu]\otimes[\lambda])^{\otimes 2}). \end{aligned}$$

Theorem 7.4.3 ([**CcW18**]). *With the notations above,*

$$\begin{aligned} &\frac{1}{\dim[\mu]\dim[\nu]}\left\|\begin{bmatrix}\alpha\beta\gamma\\ \mu\nu\\ \lambda\end{bmatrix}\right\|_1^2 \\ &\quad= \dim[\alpha]\dim[\beta]\dim[\gamma]\dim[\lambda]\dim[\mu]\dim[\nu]\operatorname{trace}(P^{\alpha\beta\gamma\lambda\mu\nu}). \end{aligned}$$

See [**CcW18**] for the proof.

Corollary 7.4.4. *With the notations above,*

$$\begin{aligned} \frac{1}{\dim[\mu]\dim[\nu]}\left\|\begin{bmatrix}\alpha\beta\gamma\\ \mu\nu\\ \lambda\end{bmatrix}\right\|_1^2 &= \frac{1}{\dim[\beta]\dim[\lambda]}\left\|\begin{bmatrix}\alpha\mu\gamma\\ \beta\lambda\\ \nu\end{bmatrix}\right\|_1^2 \\ &= \frac{1}{\dim[\alpha]\dim[\gamma]}\left\|\begin{bmatrix}\mu\beta\nu\\ \alpha\gamma\\ \lambda\end{bmatrix}\right\|_1^2. \end{aligned} \tag{7.4.4}$$

7.4.3. Proof of strong subadditivity via recoupling coefficients. Let $\rho_{ABC} \in \mathcal{D}(\mathcal{H}_A \otimes \mathcal{H}_B \otimes \mathcal{H}_C)$ and let $(\alpha_d, \dots, \lambda_d)$ be a sequence as in Theorem 7.4.2. Rewrite the first equality of equation (7.4.4) as

$$\left\|\begin{bmatrix} \alpha\beta\gamma \\ \mu\nu \\ \lambda \end{bmatrix}\right\|_1^2 = \frac{\dim[\mu]\dim[\nu]}{\dim[\beta]\dim[\lambda]} \left\|\begin{bmatrix} \alpha\mu\gamma \\ \beta\lambda \\ \nu \end{bmatrix}\right\|_1^2 .$$

The poly(d) upper bound on the square norms of the recoupling coefficients and Theorem 7.4.2 imply

$$\frac{\dim[\mu_d]\dim[\nu_d]}{\dim[\beta_d]\dim[\lambda_d]} > \frac{1}{\text{poly}(d)}.$$

Taking logs of both sides and letting d be sufficiently large, Lemma 7.1.24 implies

$$H(\bar{\mu}_d) + H(\bar{\nu}_d) - H(\bar{\beta}_d) - H(\bar{\lambda}_d) > -\frac{1}{d} C \log\left(\frac{1}{d}\right),$$

for some constant $C > 0$, and taking the limit of the sequence, we obtain

$$H(\rho_{AB}) + H(\rho_{BC}) - H(\rho_B) - H(\rho_{ABC}) \geq 0$$

which is (6.3.4), equivalent to the other statements of strong subadditivity.

7.5. Proof that $\text{Kron}_{m,n,k}$ is finitely generated and monotonicity of Kronecker coefficients

This section assumes definitions and results from commutative algebra and representation theory.

7.5.1. Positively graded Noetherian rings are finitely generated. A ring R is *Noetherian* if any ascending chain of ideals $I_1 \subseteq I_2 \subseteq \cdots$ must stabilize. Quotient rings of Noetherian rings are also Noetherian as one can lift any chain of ideals to the original ring.

Proposition 7.5.1. *A Noetherian $\mathbb{Z}_{\geq 0}$-graded ring R, with each graded component finite dimensional, is finitely generated.*

Proof. Let I_k be the ideal generated by $R_{\leq k}$. This is an increasing sequence of ideals that must stabilize and that limits to R. □

7.5.2. Finite generation of $\mathbb{C}[V]^G$. Let $V = \mathbb{C}^n$ be a module for a reductive group G, let $\mathbb{C}[V]$ denote the algebra of all polynomial functions on V, and let $\mathbb{C}[V]^G$ denote the submodule of G-invariants, which is a subalgebra of $\mathbb{C}[V]$.

Theorem 7.5.2. *Let G be reductive and let V be a G-module. Then $\mathbb{C}[V]^G$ is a finitely generated algebra.*

Proof. Consider the projection operator $\mathrm{Proj}_{\mathrm{triv}} : \mathbb{C}[V] \to \mathbb{C}[V]^G$. This is a $\mathbb{C}[V]^G$-linear map; i.e., for all $a \in \mathbb{C}[V]^G$ and $b \in \mathbb{C}[V]$, $\mathrm{Proj}_{\mathrm{triv}}(ab) = a\,\mathrm{Proj}_{\mathrm{triv}}(b)$.

Claim: $\mathbb{C}[V]^G$ is Noetherian. Were $\mathbb{C}[V]^G$ a quotient ring of $\mathbb{C}[V]$ this would be immediate. Since G is reductive, it may be realized as a quotient module. The following argument shows this will be enough to conclude. Say $I_1 \subset I_2 \subset \cdots \subset \mathbb{C}[V]^G$ is a chain of ideals. Consider the associated chain of ideals $J_k := I_k\mathbb{C}[V]$, which must stabilize as $\mathbb{C}[V]$ is Noetherian. The ideal J_k is G-stable and $J_k^G = I_k$. We conclude $I_1 \subset I_2 \subset \cdots$ must stabilize as well.

Now consider $\mathbb{C}[V]$ as a graded ring $\mathrm{Sym}(V^*) = \oplus_j S^j V^*$. This grading induces a $\mathbb{Z}_{\geq 0}$-grading of $\mathbb{C}[V]^G$, so we conclude finite generation by Proposition 7.5.1. □

7.5.3. Finite generation of $\mathbb{C}[V]^{N_G}$. For a reductive group G, one may always find a maximal unipotent subgroup N_G. In our case, for $G = \mathrm{GL}_m$, one may take N_G to be the upper triangular matrices with 1's on the diagonal.

The *algebraic Peter-Weyl theorem* (see, e.g., [**Lan17**, Thm. 8.6.6.1]) says that when G is a reductive algebraic group, one has the following isomorphism of $G \times G$-modules:

$$\mathbb{C}[G] = \bigoplus_{\lambda \in \Lambda_G^+} V_\lambda \otimes V_\lambda^*,$$

where Λ_G^+ is a set indexing the finite-dimensional irreducible representations of G, and $G \times G$ acts on the left-hand side by $((g_1, g_2)f)(h) = f(g_1 h g_2{}^{-1})$ and on the right-hand side the first G acts on V_λ and the second G on V_λ^*.

Theorem 7.5.3. *Let G be a connected reductive group, let V be a G-module, and let $N_G \subset G$ be a maximal unipotent subgroup. Then $\mathbb{C}[V]^{N_G}$ is a finitely generated algebra.*

Proof. By the algebraic Peter-Weyl theorem and the theorem of the highest weight,

$$\mathbb{C}[G]^{N_G} = \bigoplus_{\lambda \in \Lambda_G^+} V_\lambda \otimes (V_\lambda^*)^{N_G} = \bigoplus_{\lambda \in \Lambda_G^+} V_\lambda^{\oplus \dim (V_\lambda^*)^{N_G}} = \bigoplus_{\lambda \in \Lambda_G^+} V_\lambda. \tag{7.5.1}$$

We conclude the ring $\mathbb{C}[G]^{N_G}$ is finitely generated because the cone of weights in the Weyl chamber is finitely generated. Finally to see $\mathbb{C}[V]^{N_G}$ is finitely generated, observe that for any G-module M (in particular $M = \mathbb{C}[V]$), by Schur's lemma one has

$$M^{N_G} \cong (M \otimes \mathbb{C}[G]^{N_G})^G.$$ □

Proof of Theorem 7.3.15. We have, for $G = \mathrm{GL}_m \times \mathrm{GL}_n \times \mathrm{GL}_k$, an isomorphism of semi-groups $\mathrm{Kron}_{m,n,k} \cong \mathbb{C}[\mathbb{C}^m \otimes \mathbb{C}^n \otimes \mathbb{C}^k]^{N_{m,n,k}}$. □

7.5.4. Monotonicity of Kronecker coefficients. We may interpret $\mathbb{C}[G]^{N_G}$ in terms of spaces of sections of line bundles on the flag variety G/B. In the case $G = \mathrm{GL}(V)$, this is the variety of complete flags in V. Since the multiplication of two sections of a line bundle is nonzero, one obtains in the case of Kronecker coefficients the following monotonicity result:

Proposition 7.5.4 ([**Man15b**]). *For any pair of triples of partitions* (π, μ, ν), (α, β, γ),

$$k_{\pi+\alpha,\mu+\beta,\nu+\gamma} > k_{\pi\mu\nu}.$$

Appendix A

Algebra and linear algebra

A.1. Groups and rings

Let S be a set; a *binary operation* on S is a map $f : S \times S \to S$. One often writes $f(x, y) = x * y$.

Definition A.1.1. A *group* is a set G with a binary operation $*$ such that:

(1) for all $x, y, z \in G$, $x * (y * z) = (x * y) * z$ (associativity),

(2) there exists an identity element $e \in G$ such that $e * x = x * e = x$ for all $x \in G$, and

(3) for all $x \in G$, there exists an inverse $x^{-1} \in G$ such that $x * x^{-1} = x^{-1} * x = e$.

Examples of groups:

(1) $(G, *) = (\mathbb{Z}, +)$.

(2) The *general linear group* $\mathrm{GL}_n(\mathbb{C}) := \{X \in \mathrm{Mat}_{n\times n}(\mathbb{C}) \mid \det(X) \neq 0\}$, with operation matrix multiplication.

(3) The *unitary group* $\mathbf{U}(n) := \{X \in \mathrm{GL}_n(\mathbb{C}) \mid X^{-1} = \bar{X}^{\mathbf{t}}\}$, with operation matrix multiplication.

(4) Let $\mathbb{Z}/N\mathbb{Z}$ denote the set of equivalence classes in the integers defined by remainder under division by N. This is a group with operation addition mod N.

(5) Let $(\mathbb{Z}/N\mathbb{Z})^*$ denote the set of elements of $\mathbb{Z}/N\mathbb{Z}$ with multiplicative inverses. This is a group with the binary operation being the multiplication inherited from multiplication of the integers: $[x]*[y] = [x*y]$.

Exercise A.1.2. Show that $[x] * [y]$ in $(\mathbb{Z}/N\mathbb{Z})^*$ is well-defined.

Definition A.1.3. A group G is *abelian* if $x * y = y * x$ for all $x, y \in G$. An abelian group is *cyclic* if it is generated by a single element.

Examples of abelian groups: $(\mathbb{Z}, +)$ (which is also cyclic, generated by 1), $(\mathbb{Z}/N\mathbb{Z})^*$, $\mathrm{GL}_1(\mathbb{C})$, $\mathbf{U}(1)$. Nonexamples: $\mathrm{GL}_n(\mathbb{C})$, $\mathbf{U}(n)$ when $n > 1$.

We will see (Theorem 3.3.8) that $(\mathbb{Z}/p\mathbb{Z})^*$ is cyclic when p is prime.

Definition A.1.4. A *ring* is a set R with two binary operations, often denoted $+$ and $*$, such that:

(1) $(R, +)$ is an abelian group, with identity element usually denoted 0.

(2) $*$ is associative and has a unit element (one says it is *unital*), usually denoted 1 or 1_R.

(3) (Compatibility) For all $x, y, z \in R$, $(x + y) * z = x * z + y * z$ and $z * (x + y) = z * x + z * y$.

Examples of rings:

(1) $(\mathbb{Z}, +, *)$, $(\mathbb{Z}/N\mathbb{Z}, +, *)$ where both operations for $\mathbb{Z}/N\mathbb{Z}$ are inherited from the operations on $\mathbb{Z}$.

(2) $\mathbb{Z}[x]$: the polynomials in one variable with integer coefficients.

(3) $\mathrm{Mat}_{n \times n}$: the ring of $n \times n$-matrices.

(4) $\mathbb{C}[z]/(z^N - 1)$: the quotient of the ring of polynomials with complex coefficients by the relation $[z^N] = [1]$.

To illustrate the last example, consider $\mathbb{C}[z]/(z^3 - 1)$. Any element of this ring may be written as $[az^2 + bz + c]$ for some $a, b, c \in \mathbb{C}$. Multiplication is given by

$$\begin{aligned}
&[az^2 + bz + c] * [a'z^2 + b'z + c'] \\
&\quad = [aa'z^4 + (a'b + ab')z^3 + (ac' + bb' + a'c)z^2 + (bc' + b'c)z + cc'] \\
&\quad = [(ac' + bb' + a'c)z^2 + (aa' + bc' + b'c)z + (cc' + a'b + ab')].
\end{aligned}$$

Definition A.1.5. Let G, H be groups. A map $f : G \to H$ is a *group homomorphism* if for all $x, y \in G$, $f(x * y) = f(x) * f(y)$ where the first $*$ is in G and the second in H.

Let R, S be rings. A map $f : R \to S$ is a *ring homomorphism* if for all $x, y \in R$:

(1) $f(1_R) = 1_S$.
(2) $f(x + y) = f(x) + f(y)$.
(3) $f(x * y) = f(x) * f(y)$.

A.2. Linear and multilinear algebra

The reader is assumed to be familiar with terms such as vector space, linear map, eigenvalue, etc.

Definition A.2.1. For a finite-dimensional vector space V over a field $\mathbb{F}$, define the *dual space*

$$V^* := \{f : V \to \mathbb{F} \mid f \text{ is linear}\}.$$

If $V = \mathbb{F}^n$ is the space of column vectors, then V^* may be identified with the space of row vectors. Given $\alpha = (a_1, \ldots, a_n) \in V^*$ and $v = (c_1, \ldots, c_n)^{\mathbf{t}} \in V$, $\alpha(v) = \sum_j a_j c_j$ is just the result of row-column multiplication.

There is a canonical isomorphism $V \to (V^*)^*$, given by $v \mapsto f_v$ where $f_v(\alpha) := \alpha(v)$. Thus we may also think of V as the space of linear maps $V^* \to \mathbb{F}$. Following physics convention, we often denote elements of V by $|v\rangle$ and elements of V^* by $\langle\phi|$. Let $\mathrm{End}(V)$ denote the space of linear maps $V \to V$, the *endomorphisms* of V.

Let V, W be finite-dimensional vector spaces of dimensions n, m. Given $v \in V$ and $\beta \in W^*$, define a rank one linear map $v \otimes \beta : W \to V$ by $(v \otimes \beta)(w) := \beta(w)v$. In bases, $v \otimes \beta$ is the product of the column vector v with the row vector β to get the rank one $n \times m$-matrix $v\beta$. The image of $v \otimes \beta$ is the line through v and its kernel is $\beta^\perp := \{w \in W \mid \beta(w) = 0\}$. Geometrically, $\beta^\perp$ is a hyperplane in W. Write $\mathrm{Hom}(W, V)$ for the space of all linear maps $W \to V$. If $v_1, \ldots, v_n$ is a basis of V and $\beta_1, \ldots, \beta_m$ is a basis of W^*, then the elements $v_i \otimes \beta_s$, $1 \le i \le n$, $1 \le s \le m$, form a basis of $\mathrm{Hom}(W, V)$. Indeed, the matrix corresponding to $v_i \otimes \beta_s$ will have a 1 in the (i, s) slot and zeros elsewhere.

We may also view $v \otimes \beta$ as a bilinear map $V^* \times W \to \mathbb{F}$; namely $v \otimes \beta(\alpha, w) = \alpha(v)\beta(w)$. To make the notation more symmetric, write $W^* = U$. Let $V \otimes U$ denote the space of bilinear maps $V^* \times U^* \to \mathbb{F}$.

By the discussion above, we have a canonical isomorphism $V \otimes V^* \cong \mathrm{End}(V)$.

In summary and generalizing:

Definition A.2.2. Define the *tensor product* $V \otimes W$ of vector spaces V and W to be the space of bilinear maps $V^* \times W^* \to \mathbb{F}$. For a collection of vector spaces $V_1, \ldots, V_m$, $V_1 \otimes \cdots \otimes V_m$ is the space of m-linear maps $V_1^* \times \cdots \times V_m^* \to \mathbb{F}$.

If $v_k \in V_k$, $1 \leq k \leq m$, define the *tensor product of vectors* $v_1 \otimes \cdots \otimes v_m$ to be the m-linear map $v_1 \otimes \cdots \otimes v_m(\alpha_1, \ldots, \alpha_m) = \alpha_1(v_1) \cdots \alpha_m(v_m)$.

If we work in bases and if $\dim V_j = \mathbf{v}_j$, then $V_1 \otimes V_2$ is the space of $\mathbf{v}_1 \times \mathbf{v}_2$-matrices and $V_1 \otimes V_2 \otimes V_3$ may be visualized as the space of $\mathbf{v}_1 \times \mathbf{v}_2 \times \mathbf{v}_3$-"three-dimensional matrices".

Definition A.2.3. Given a linear map $f : V \to W$, we may define a second linear map $f^{\mathbf{t}} : W^* \to V^*$ by, for $\beta \in W^*$, $f^{\mathbf{t}}(\beta)(v) = \beta(f(v))$. This is the coordinate-free definition of the *transpose*.

One may also use $f : V \to W$ to define a bilinear map $W^* \times V \to \mathbb{F}$, by $(\beta, v) \mapsto \beta(f(v))$, which extends to a linear map $W^* \otimes V \to \mathbb{C}$.

Consider a 2×3-matrix

$$\begin{pmatrix} a & b & c \\ d & e & f \end{pmatrix}$$

and its roles, respectively, as a linear map $\mathbb{F}^3 \to \mathbb{F}^2$, a linear map $\mathbb{F}^{2*} \to \mathbb{F}^{3*}$, and a bilinear map $\mathbb{F}^{2*} \times \mathbb{F}^3 \to \mathbb{F}$:

$$\begin{pmatrix} a & b & c \\ d & e & f \end{pmatrix} \begin{pmatrix} x \\ y \\ z \end{pmatrix} = \begin{pmatrix} ax + by + cz \\ dx + ey + fz \end{pmatrix},$$

$$\begin{pmatrix} s & t \end{pmatrix} \begin{pmatrix} a & b & c \\ d & e & f \end{pmatrix} = (sa + td, sb + te, sc + tf),$$

$$\begin{pmatrix} s & t \end{pmatrix} \begin{pmatrix} a & b & c \\ d & e & f \end{pmatrix} \begin{pmatrix} x \\ y \\ z \end{pmatrix} = sax + tdx + sby + tey + scz + tfz.$$

Definition A.2.4. A *Lie algebra* is a vector space $\mathfrak{g}$ equipped with a skew-symmetric multiplication, traditionally denoted $(X, Y) \mapsto [X, Y]$, which satisfies the *Jacobi identity*: for $X, Y, Z \in \mathfrak{g}$,

$$[Z, [X, Y]] = [[Z, X], Y] + [X, [Z, Y]].$$

The Jacobi identity says $[Z, \cdot] : \mathfrak{g} \to \mathfrak{g}$ is a derivation.

The most classical example of a Lie algebra is $\mathfrak{gl}_n$, where

$$[X, Y] = XY - YX$$

is the *commutator*.

Appendix B

Probability

B.1. Probability distributions and marginals

Definition B.1.1. Let $\mathcal{X} = \{a_1, a_2, \dots\}$ be a countable set (in this book $\mathcal{X}$ will usually be finite). A function $P : \mathcal{X} \to [0,1]$ such that $\sum_j P(a_j) = 1$ is called a *(discrete) probability distribution* on $\mathcal{X}$. When $\mathcal{X}$ is equipped with a probability distribution, a function $X : \mathcal{X} \to \mathbb{R}$ is called a *discrete random variable.*

A discrete random variable defines a probability distribution with countable support on $\mathbb{R}$ by

$$P_X(z) = \sum_{j|X(a_j)=z} P(a_j).$$

In particular, $P_X(z) = 0$ if $z \notin X(\mathcal{X})$. Random variables X, Y define a probability distribution $P_{X,Y}(x,y)$ with discrete support on $\mathbb{R} \times \mathbb{R}$, and n random variables define a probability distribution with discrete support on $\mathbb{R}^n$. If $f : \mathbb{R} \to \mathbb{R}$ is a function, then $f \circ X$ is also a random variable.

Definition B.1.2. Let P be a probability distribution on $\mathcal{X} \times \mathcal{Y}$. Define the *marginals* by

$$P_{\mathcal{X}}(x) = \sum_{y \in \mathcal{Y}} P(x,y) \quad \text{and} \quad P_{\mathcal{Y}}(y) = \sum_{x \in \mathcal{X}} P(x,y),$$

which are probability distributions on $\mathcal{X}$, $\mathcal{Y}$, respectively.

B.2. Expectation and variance

Definition B.2.1. The *expectation* (or *average*) of a random variable X on a countable set $\mathcal{X}$ equipped with a probability distribution p is

$$E[X] = E_p[X] = \sum_{x \in \mathcal{X}} X(x)p(x). \tag{B.2.1}$$

Note that $E[X] \in [-\infty, \infty]$.

For example, if $\mathcal{X} = \{H, T\}$ are the possible outcomes of flipping a biased coin which lands heads (H) with probability p and tails (T) with probability $1 - p$ and if $X(H) = 1, X(T) = -1$, then $E[X] = 2p - 1$, which is zero if the coin is fair.

Notation. $\Pr(Z)$ denotes the probability, with respect to some understood distribution, that some element of the subset $Z \subset \mathcal{X}$ occurs. In this context, the subset Z is called an *event*.

Proposition B.2.2 (Markov's inequality). *Let X be a random variable on $\mathcal{X}$ such that $X(x) \geq 0$ for all $x \in \mathcal{X}$. Then for all $a > 0$,*

$$\Pr(X \geq a) \leq \frac{E[X]}{a}. \tag{B.2.2}$$

Proof. Define a new random variable Y_a by

$$Y_a(x) = \begin{Bmatrix} 0, & X(x) < a \\ a, & X(x) \geq a \end{Bmatrix}.$$

Since for all x, $Y_a(x) \leq X(x)$, we have $E[Y_a] \leq E[X]$. But

$$E[Y_a] = a\Pr(X \geq a) + 0\Pr(X < a).$$ □

For example, if X satisfies the hypotheses of Markov's inequality, then $\Pr(X \geq NE[X]) \leq \frac{1}{N}$.

Exercise B.2.3. Show that Markov's inequality is sharp; i.e., write down an example where equality occurs in (B.2.2).

To get better bounds, one can utilize more information. A first measurement of how far one can expect to be from the expectation is the *variance*:

Definition B.2.4. The *variance* of X is

$$\mathrm{var}(X) := E[(X - E[X])^2]. \tag{B.2.3}$$

Exercise B.2.5. Show that $\mathrm{var}(X) = E[X^2] - E[X]^2$.

Proposition B.2.6 (Chebyshev's inequality). *Let X be a random variable. For all $c > 0$*

$$\Pr(|X - E[X]| \geq c) \leq \frac{\mathrm{var}(X)}{c^2}. \tag{B.2.4}$$

Proof. Define the random variable $Y(x) = (X(x) - E[X])^2$. Since $Y(x) \geq 0$ for all x, Markov's inequality implies

$$\Pr(Y \geq c^2) \leq \frac{E[Y]}{c^2};$$

i.e.,

$$\Pr((X - E[X])^2 \geq c^2) \leq \frac{E[Y]}{c^2}.$$

Finally, since $c > 0$, $(X - E[X])^2 \geq c^2$ is the same as $|X - E[X]| \geq c$. □

Exercise B.2.7. Show that Chebyshev's inequality is sharp; i.e., write down an example where equality occurs in (B.2.4).

Often one deals with the square root of the variance of a random variable:

Definition B.2.8. The *standard deviation* of a random variable X is

$$\sigma(X) := \sqrt{\operatorname{var}(X)}.$$

The standard deviation and variance measure the failure of the corresponding probability distribution to be concentrated at a point, i.e., the failure of the induced probability distribution to have a certain outcome.

B.3. The law of large numbers

Definition B.3.1. A probability distribution $p_{\mathcal{X}\times\mathcal{Y}}$ induces *independent* marginal distributions $p_{\mathcal{X}}, p_{\mathcal{Y}}$ if for all $x \in \mathcal{X}$ and $y \in \mathcal{Y}$, $p_{\mathcal{X}\times\mathcal{Y}}(x, y) = p_{\mathcal{X}}(x)p_{\mathcal{Y}}(y)$.

Random variables X, Y on $\mathcal{X}$ are said to be *independent* if for all $x_1, x_2 \in \mathbb{R}$,

$$p_{X,Y}(x_1, x_2) = p_X(x_1)p_Y(x_2)$$

and one extends the notion to random variables $X_1, \dots, X_m$. They are *identically distributed* if they define the same probability distributions. Write $X_1, \dots, X_m$ are *iid* if they are independent and identically distributed.

A typical situation is to define random variables X_j, where $X_j = 1$ if the outcome of the j-th toss of a coin is heads and $X_j = -1$ if the outcome of the j-th toss is tails. Then the X_j are iid.

The *law of large numbers* implies that the name "expectation" is reasonable; that is, if one makes repeated experiments (e.g., as with the coin flips above) and averages the outcomes, the averages limit to the expectation.

More precisely:

Proposition B.3.2 (Weak law of large numbers). *For $X, X_1, X_2, \dots$ independent identically distributed random variables with finite expectation and variance and for any $\epsilon > 0$,*

$$\lim_{d\to\infty} \Pr\left(\left|\frac{X_1 + \cdots + X_d}{d} - E[X]\right| \geq \epsilon\right) = 0. \tag{B.3.1}$$

Proof. Define the random variable $M_d := \frac{1}{d}(X_1 + \cdots + X_d)$ and note that $E[M_d] = E[X]$ and $\mathrm{var}(M_d) = \frac{1}{d}\,\mathrm{var}(X)$. Chebyshev's inequality says

$$\Pr(|M_d - E[X]| \geq \epsilon) \leq \frac{\mathrm{var}(M_d)}{\epsilon^2} = \frac{\mathrm{var}(X)}{d\epsilon^2}$$

and the last term goes to zero as $d \to \infty$. □

We may rephrase (B.3.1) as follows: for all $\epsilon, \delta > 0$, there exists a d_0 such that for all $d \geq d_0$

$$\Pr\left(\left|\frac{X_1 + \cdots + X_d}{d} - E[X]\right| \geq \epsilon\right) < \delta. \tag{B.3.2}$$

B.4. Joint probability distributions and conditional probability

Definition B.4.1. Let $p_{\mathcal{X}\times\mathcal{Y}}$ be a probability distribution on $\mathcal{X} \times \mathcal{Y}$. The *conditional probability* of x occurring given knowledge that y occurs (assuming $p_{\mathcal{Y}}(y) > 0$) is

$$\begin{aligned} p_{\mathcal{X}|\mathcal{Y}}(x|y) &= \frac{p_{\mathcal{X}\times\mathcal{Y}}(x,y)}{p_{\mathcal{Y}}(y)} \\ &= \frac{p_{\mathcal{X}\times\mathcal{Y}}(x,y)}{\sum_{a\in\mathcal{X}} p_{\mathcal{X}\times\mathcal{Y}}(a,y)}. \end{aligned}$$

Note that $p_{\mathcal{X}|\mathcal{Y}}(\cdot|y)$ is a probability distribution on $\mathcal{X}$ and that one can recover $p_{\mathcal{X}\times\mathcal{Y}}$ from $p_{\mathcal{X}|\mathcal{Y}}$ and $p_{\mathcal{Y}}$.

One may similarly define the conditional probability of events.

Conditional probability is often used to compute the probability of an event via a "divide and conquer" strategy: let $\mathcal{X} = \bigsqcup_s B_s$ be a disjoint union; then for $A \subset \mathcal{X}$,

$$\Pr(A) = \sum_s \Pr(A|B_s)\Pr(B_s),$$

which is often called the *law of total probability*.

Exercise B.4.2. Given $x \in \mathcal{X}$ and $\mathcal{X} = \bigsqcup_s B_s$, what is $\Pr(x|B_s)$ in terms of $\Pr(x)$ and $\Pr(B_s)$?

Hints and answers to selected exercises

Chapter 1.

1.2.8. First do the case that one of the $n \times n$-matrices has distinct eigenvalues. Note that if v is an eigenvector for A with eigenvalue λ, i.e., $Av = \lambda v$, then $ABv = BAv = \lambda Bv$ so Bv is also an eigenvector for A with eigenvalue λ. Finally do the general case either using a continuity argument or letting λ be an eigenvalue of A with eigenspace spanned by $v_1, \ldots, v_k$; then show that B preserves the space $\langle v_1, \ldots, v_k \rangle$ so it must have an eigenvector inside it and work by induction.

1.3.7. If the function takes one only on the input 10011, then take $x_1 \wedge (\neg x_2) \wedge (\neg x_3) \wedge x_4 \wedge x_5$.

1.3.8. If the function takes one on the inputs 10011 and 00001, then take $(x_1 \wedge (\neg x_2) \wedge (\neg x_3) \wedge x_4 \wedge x_5) \vee ((\neg x_1) \wedge (\neg x_2) \wedge (\neg x_3) \wedge (\neg x_4) \wedge x_5)$.

1.3.12. First note that $ab \bmod N = (a \bmod N)(b \bmod N) \bmod N$; so if we have a fast way to compute $c \bmod N$ for $c < N^2$, we only need to multiply numbers less than N. For the case $k = 2^\ell$, we have

$$a^4 \bmod N = (a^2 \bmod N)(a^2 \bmod N) \bmod N,$$

etc.

1.5.1. $\mathrm{DFT}_N \otimes \mathrm{DFT}_M$; i.e., $|js\rangle \mapsto \sum_k \sum_t \omega_N^{jk} \omega_M^{st} |kt\rangle$.

1.5.2. $\begin{pmatrix} 1 & 1 \\ 1 & -1 \end{pmatrix}^{\otimes k}$.

1.6.2. Say we have three elements x, y, z and we would like to have the uniform distribution on them. Consider the eight-element set $x_1, x_2, x_3, y_1, y_2, y_3, z_1, z_2$ and give it the uniform distribution. Then if an x_j is drawn, choose x, etc.

Chapter 2.

2.1.12. Write $A_1 := A|_{v_1^\perp}$ (the restriction of A to $v_1^\perp$). Note that A_1 must have an eigenvector and that it is unitary. Repeat $n-1$ times.

2.3.11. Differentiate $U(t)v(t)$ where $v(t)$ is an eigenvector of $U(t)$ with eigenvalue $\lambda(t)$ and $U(0) = \text{Id}$, so $\lambda(0) = 1$.

2.3.12. Use Exercise 2.1.12.

2.3.13. It is sufficient to work in bases, i.e., with matrices. First prove the case X is diagonal, then the case X is diagonalizable. Finally either write $X = X_s + X_n$ as the sum of a diagonalizable matrix and a nilpotent matrix (Jordan normal form) or argue that the set of diagonalizable matrices is a dense open subset in the space of all matrices.

2.3.15. First consider the case X is diagonal. Then use that Hermitian matrices are diagonalizable.

2.6.4. $\cos^2(\frac{\pi}{8})$.

Chapter 3.

3.1.1. It suffices to do the case $v = \begin{pmatrix} 1 \\ 0 \end{pmatrix}$.

3.1.5. Use the approximation that for θ small, $\sin(\theta) \simeq \theta$.

3.1.7. The probability of measuring $|a\rangle$ after k applications of R is

$$(\sin(2k+1)\sin^{-1}(2^{-\frac{n}{2}}))^2 \quad \text{and} \quad \sin^{-1}(2^{-\frac{n}{2}}) \sim 2^{-\frac{n}{2}}.$$

3.2.4. This is just asking that $2n$ vectors in $\mathbb{F}_2^{n-1}$ selected uniformly at random will span the space. There are 2^k vectors in $\mathbb{F}_2^k$, so if we have already selected k independent vectors, the odds that a $(k+1)$-st selected uniformly at random will be independent of the others is $1 - \frac{2^k}{2^{n-1}}$.

3.3.3.

$$\begin{pmatrix} F_n \\ F_{n-1} \end{pmatrix} = \begin{pmatrix} 1 & 1 \\ 1 & 0 \end{pmatrix} \begin{pmatrix} F_{n-1} \\ F_{n-2} \end{pmatrix} \quad \text{and} \quad \begin{pmatrix} r_n \\ r_{n+1} \end{pmatrix} = \begin{pmatrix} \alpha_{n+2} & 1 \\ 1 & 0 \end{pmatrix} \begin{pmatrix} r_{n+1} \\ r_{n+2} \end{pmatrix}.$$

3.3.5. There exist $s, t \in \mathbb{Z}$ such that $1 = sa + tm$. Then $x = sb$ is a solution. Say y is another solution. Then $0 = b - b = a(y - sb) \equiv 0 \bmod m$.

3.3.6. There exist s, t such that $1 = sm + tn$. Take $x = cms + bnt$. Say y is another solution; then both m and n divide $x - y$.

3.3.13. Prove it in the form $x^p \equiv x \bmod p$. Use induction: the case $x = 1$ is easy, and use Exercise 3.3.12 to prove the induction step.

3.3.15. The only noninvertible elements are the multiples of p, and there are p^{k-1} of them.

3.3.17. In going from the case $k = 2$ to $k = 3$.

3.4.1. Write $a = xd$, $N = yd$. Were $a^{N-1} \equiv 1 \bmod N$ we would have, for some $q \geq 1$, $d^q(x^{N-1}d^{N-1-q} - y^q) = 1$, a contradiction.

3.4.6. $a^{N-1} \equiv p^{c-1} + 1 \bmod N$ and $p^{c-1} + 1 \not\equiv 1 \bmod N$.

3.4.8. Say p divides N, so $N = ap^m$ for some a, m. Then

$$\binom{N}{p} = \frac{ap^m}{p} \frac{(ap^m - 1)!}{(p-1)!\,(ap^m - p)!}$$

and note the second factor is not divisible by p

3.6.1. There are many ways to prove this. Here is an elementary one: first suppose $G = \mathbb{Z}/N\mathbb{Z} = \{0, \dots, N-1\}$ is cyclic of order $N = 2m$. Let $x \in G$ be of odd order k. Claim: $x + 1$ is of even order, which will be enough to conclude in this case. We have that $2m$ divides kx. Say $x + 1$ is also of odd order k'; then $2m$ divides $k'(x + 1)$ and thus $2m$ divides $(k + k')x + k'$. But $(k + k')x$ is even and k' is odd, so $(k + k')x + k'$ is odd, a contradiction.

Any finite abelian group is a product of cyclic groups. Write $G = \mathbb{Z}/N_1\mathbb{Z} \times \cdots \times \mathbb{Z}/N_h\mathbb{Z}$, and suppose $N_1 = 2m$ is even. Let $(x_1, x_2, \dots, x_h) \in G$ be of odd order, so each x_i is of odd order. Using the special case, we have that $(x_1 + 1, x_2, \dots, x_h)$ is of even order.

3.6.5. Write $x = \lfloor x \rfloor + \{x\}$. If $\{x\} < \frac{1}{2}$, then $2x = 2\lfloor x \rfloor + \{2x\}$ and therefore $\lfloor 2x \rfloor = 2\lfloor x \rfloor$. If $\{x\} \geq \frac{1}{2}$, then $\lfloor 2x \rfloor = \lfloor (2\lfloor x \rfloor) \rfloor + 1$.

3.6.14. Recall that a, b are relatively prime if and only if there exist integers s, t such that $as + bt = 1$.

3.6.21. Write $(-1)^n = p_n q_{n-1} - q_n p_{n-1}$.

Chapter 4.

4.3.3. Were s a continuous variable, you could just solve the ODE. However, the monotonicity (3) is almost as good: fix t and consider increasing integers n and $m(n)$ such that $2^{m(n)} \leq t^n < 2^{m(n)+1}$. Take logs and divide by $n \log(s)$ to obtain

$$\left| \frac{m}{n} - \frac{\log(t)}{\log(s)} \right| < \epsilon$$

for an arbitrarily small $\epsilon = \epsilon(n)$. Now use the monotonicity to get a second estimate involving A.

4.5.13. Show that the log of the quantity goes to zero using l'Hôpital's rule.

Chapter 5.

5.1.3.

$$\begin{aligned}\Pr(\operatorname{span}\{\mathcal{M}_1,\mathcal{M}_2\}) &= 1,\\ \Pr(\mathcal{M}_1) &= |\alpha|^2,\\ \Pr(\mathcal{M}_2) &= \frac{1}{2}|\alpha+\beta|^2,\\ \Pr(\mathcal{M}_1\cap\mathcal{M}_2) &= 0.\end{aligned}$$

5.1.5. First try the case $\mathcal{M}_1, \mathcal{M}_2$ are orthogonal, then the case $\mathcal{M}_1 \subset \mathcal{M}_2$. The general commuting case is a combination of these.

5.1.9. Use that any $X \in \operatorname{End}(\mathcal{H})$ may be uniquely written as a sum of a Hermitian and an anti-Hermitian (i.e., i times a Hermitian) operator.

5.1.14. Use that $\operatorname{Proj}_{\mathcal{M}}^2 = \operatorname{Proj}_{\mathcal{M}} = \operatorname{Proj}_{\mathcal{M}}^{\dagger}$ and $\operatorname{trace}(AB) = \operatorname{trace}(BA)$.

5.2.1. The trace.

5.2.8. Show that there exists an operator whose eigenspaces are the same as for ρ and whose eigenvalues are the positive square roots of the eigenvalues of ρ. Or see §6.1.2 for the proof of a similar statement.

5.2.14. A POVM on $\mathcal{H}$ may be interpreted as a projective measurement on some $\mathcal{H}\otimes\mathcal{H}'$ with $\dim \mathcal{H}' = \ell$, where ℓ is the number of Q_s's.

5.2.15. Use Exercise 2.3.12.

5.2.18. Use $\operatorname{trace}(Z^{\dagger}) = \overline{\operatorname{trace}(Z)}$ to show $\operatorname{trace}([X,Y]\rho) = -\overline{\operatorname{trace}([X,Y]\rho)}$.

5.2.19. Use $\operatorname{trace}(Z^{\dagger}Z\rho) = \operatorname{trace}(Z\rho Z^{\dagger})$.

5.5.2. Note that $(\tau\otimes \operatorname{Id})(\sum_{ij}|ii\rangle\langle jj|) = \sum_{ij}|ji\rangle\langle ij|$, and this image operator has some eigenvalues equal to -1; e.g., $|12\rangle - |21\rangle \mapsto |21\rangle - |12\rangle$.

5.5.6. $\rho\otimes\sigma\otimes \operatorname{Id}_{\mathcal{H}_C}$ is positive as $\rho, \sigma, \operatorname{Id}_{\mathcal{H}_C}$ are each positive; acting by a unitary preserves positivity and thus an isometry also preserves positivity. Finally, the partial trace of a positive operator is positive as the eigenvalues of the new operator are sums of eigenvalues of the old. Trace preserving follows similarly.

5.5.13. Write $\sigma = \sum_s \lambda_s |s\rangle\langle s|$ and consider $\sqrt{\lambda_s}|s\rangle$.

5.5.17. We have $\operatorname{trace}(\Lambda(X)) = \operatorname{trace}(\sum_\alpha Y_\alpha X Y_\alpha^{\dagger})$. Use the cyclic invariance of the trace and the fact that $\operatorname{trace}(XZ) = \operatorname{trace}(X'Z)$ for all Z implies $X = X'$; i.e., the map $(X,Z) \mapsto \operatorname{trace}(XZ)$ is a perfect pairing $\operatorname{End}(\mathcal{H}) \times \operatorname{End}(\mathcal{H}) \to \mathbb{C}$. (In more abstract language, $\operatorname{End}(\mathcal{H})$ is naturally self-dual and the trace realizes the self-duality pairing.)

Chapter 6.

6.1.8. Use Theorem 6.1.6 and the Cauchy-Schwartz inequality.

6.2.12. Write the eigenvalue decompositions

$$\rho = \sum \lambda_j |\psi_j\rangle\langle\psi_j|, \quad \sigma = \sum \mu_k |\eta_j\rangle\langle\eta_j|.$$

Then

$$\begin{aligned}\log(\rho\otimes\sigma) &= \sum_{j,k} \log(\lambda_j\mu_k)|\psi_j\rangle\langle\psi_j|\otimes|\eta_j\rangle\langle\eta_j| \\ &= \sum_{j,k} \log(\lambda_j)|\psi_j\rangle\langle\psi_j|\otimes|\eta_j\rangle\langle\eta_j| + \sum_{j,k} \log(\mu_k)|\psi_j\rangle\langle\psi_j|\otimes|\eta_j\rangle\langle\eta_j| \\ &= \sum_{j} \log(\lambda_j)|\psi_j\rangle\langle\psi_j|\otimes \mathrm{Id}_{\mathcal{H}_B} + \sum_{k} \log(\mu_k)\,\mathrm{Id}_{\mathcal{H}_A} \otimes|\eta_j\rangle\langle\eta_j|.\end{aligned}$$

6.2.20. Imitate the proof of Proposition 6.2.8.

6.3.3. Introduce a purification ρ_{ABCR} of ρ_{ABC}, $\rho_{ABC} = \mathrm{trace}_{\mathcal{H}_R}(\rho_{ABCR})$. Then, by (6.3.3), $H(\rho_R) + H(\rho_B) \leq H(\rho_{CR}) + H(\rho_{BC})$, but since ρ_{ABCR} is pure $H(\rho_R) = H(\rho_{ABC})$ and $H(\rho_{CR}) = H(\rho_{AB})$ and we conclude.

6.3.4. Proof of (1): Considering $\mathcal{H}_B\otimes\mathcal{H}_D$ as a single space, (6.3.4) implies

$$H(\rho_{ABCD}) + H(\rho_C) \leq H(\rho_{AC}) + H(\rho_{BCD}).$$

Now add $H(\rho_D)$ to both sides and observe $H(\rho_{BCD}) + H(\rho_D) \leq H(\rho_{AC}) + H(\rho_{BD}) + H(\rho_{BC})$; substitute this in and rearrange to obtain the result.

The first assertion in (2) is just (6.3.4). The second is equivalent to

$$H(\rho_{ABC}) + H(\rho_B) + H(\rho_C) \leq H(\rho_{AB}) + H(\rho_{BC}) + H(\rho_{AC}).$$

By strong subadditivity, at least one of the inequalities $H(\rho_C) \leq H(\rho_{AC})$, $H(\rho_B) \leq H(\rho_{AB})$ must hold. Say the first does; adding it to the equation gives the result.

6.3.9. Find U that diagonalizes X and then apply permutations to average.

6.3.10. First show that $H(\rho_A||\sigma_A)=H(\rho_A\otimes\frac{1}{\dim\mathcal{H}_B}\mathrm{Id}_{\mathcal{H}_B}||\sigma_A\otimes\frac{1}{\dim\mathcal{H}_B}\mathrm{Id}_{\mathcal{H}_B})$. Then note that, by the same reasoning as in Exercise 6.3.9, one may send ρ_{AB} to $\rho_A\otimes\frac{1}{\dim\mathcal{H}_B}\mathrm{Id}_{\mathcal{H}_B}$ via a mixed unitary channel.

6.5.1. Note that $(M_q)_{st} = \sum_{\sigma\in\mathfrak{S}_n|\sigma(s)=t} q_\sigma$. Now sum over t. Similarly, sum over s.

6.5.5. Show that $x \prec y$ if and only if $\forall t \in \mathbb{R}$, $\sum_{j=1}^n \max(x_j - t, 0) \leq \sum_{j=1}^n \max(y_j - t, 0)$ and $\sum_{j=1}^n x_j = \sum_{j=1}^n y_j$.

6.5.7. Say $x \in \mathrm{conv}\{y_1, \dots, y_m\}$, so $x = \sum p_s y_s$ with $\bar{p}$ a probability distribution. Then $\alpha(x) = \sum p_s \alpha(y_s) \leq \max_s \alpha(y_s)$. In the other direction, the level sets of α are affine hyperplanes and $x \notin \mathrm{conv}\{y_1, \dots, y_m\}$ exactly says there is some affine hyperplane separating x from $\mathrm{conv}\{y_1, \dots, y_m\}$. Take α with one such hyperplane as a level set.

6.5.11. Assume wlog that the right-hand side is trace(A) as both sides are invariant under permutation matrices.

Chapter 7.

7.1.3. Consider the decomposition

$$\mathbb{C}^d = \mathbb{C}\{e_1 + e_2 + \cdots + e_m\} \oplus \{v = v_1 e_1 + \cdots + v_m e_m \mid \sum_i v_i = 0\}.$$

7.1.7. Show that the kernel and image of a G-module map are submodules.

7.1.8. Consider $\mathbb{C}\{\mathrm{Id}\}$. For the uniqueness, use Schur's lemma.

7.1.11. The symmetrization map is $v_1 \otimes v_2 \otimes v_3 \otimes v_4 \mapsto v_1 \otimes v_2 \otimes v_3 \otimes v_4 + v_2 \otimes v_1 \otimes v_3 \otimes v_4 + v_1 \otimes v_2 \otimes v_4 \otimes v_3 + v_2 \otimes v_1 \otimes v_4 \otimes v_3$ and then each term $a \otimes b \otimes c \otimes d$ is mapped to $a \otimes b \otimes c \otimes d - b \otimes a \otimes c \otimes d - a \otimes b \otimes d \otimes c + b \otimes a \otimes d \otimes c$.

7.1.12. Let $e_1, \dots, e_n$ be a basis of V; then $e_1 \otimes \cdots \otimes e_d$ will not map to zero under any projection operator. On the other hand, when $\dim V < d$, the projection onto (1^d) will be zero.

7.1.21. There are $\binom{m}{2}$ terms in (7.1.2).

7.1.23. The quantity on the left is just the product of the horizontal contributions to the hook lengths and the quantity on the right assumes the vertical contributions are as large as possible.

7.2.3. Split d into n intervals; take a Young tableau such that no column has more than one index from each interval.

7.2.5. We are projecting to the entire isotypic component so it suffices to find one Young tableau giving rise to a nonzero projection. Now construct a Young tableau-tailor made to the ordering of I.

7.3.7. We even have $\dim S^d \mathbb{C}^m = O(d^m)$. Let $\mathbb{C}^m = \mathbb{C}^M \otimes \mathbb{C}^M \otimes \mathbb{C}^M$.

Bibliography

[A99] A. Aspect, *Bell's inequality test: more ideal than ever*, Nature **398** (1999), no. 6724, 189–90.

[Aar13] S. Aaronson, *Quantum computing since Democritus*, Cambridge University Press, Cambridge, 2013, DOI 10.1017/CBO9780511979309. MR3058839

[AB09] S. Arora and B. Barak, *Computational complexity: A modern approach*, Cambridge University Press, Cambridge, 2009, DOI 10.1017/CBO9780511804090. MR2500087

[Ad89] V. I. Arnol′d, *Mathematical methods of classical mechanics*, translated from the 1974 Russian original by K. Vogtmann and A. Weinstein, corrected reprint of the second (1989) edition, Graduate Texts in Mathematics, vol. 60, [1989?]. MR1345386

[Adl78] L. Adleman, *Two theorems on random polynomial time*, 19th Annual Symposium on Foundations of Computer Science (Ann Arbor, Mich., 1978), IEEE, Long Beach, CA, 1978, pp. 75–83. MR539832

[Ahl78] L. V. Ahlfors, *Complex analysis: An introduction to the theory of analytic functions of one complex variable*, 3rd ed., International Series in Pure and Applied Mathematics, McGraw-Hill Book Co., New York, 1978. MR510197

[AJSP14] A. Peruzzo, J. McClean, P. Shadbolt, M.-H. Yung, X.-Q. Zhou, P. J. Love, A. Aspuru-Guzik, and J. L. O'Brien, *A variational eigenvalue solver on a photonic quantum processor*, Nat. Commun. **4213** (2014).

[AKS04] M. Agrawal, N. Kayal, and N. Saxena, *PRIMES is in P*, Ann. of Math. (2) **160** (2004), no. 2, 781–793, DOI 10.4007/annals.2004.160.781. MR2123939

[ARS88] R. Alicki, S. Rudnicki, and S. Sadowski, *Symmetry properties of product states for the system of N n-level atoms*, J. Math. Phys. **29** (1988), no. 5, 1158–1162, DOI 10.1063/1.527958. MR941037

[AW21] J. Alman and V. Vassilevska Williams, *A refined laser method and faster matrix multiplication*, Proceedings of the 2021 ACM-SIAM Symposium on Discrete Algorithms (SODA), [Society for Industrial and Applied Mathematics (SIAM)], Philadelphia, PA, 2021, pp. 522–539, DOI 10.1137/1.9781611976465.32. MR4262465

[BBP$^+$96] C. H. Bennett, G. Brassard, S. Popescu, B. Schumacher, J. A. Smolin, and W. K. Wootters, *Purification of noisy entanglement and faithful teleportation via noisy channels*, Phys. Rev. Lett. **76** (1996), 722–725.

[BCHW16] F. G. S. L. Brandao, M. Christandl, A. W. Harrow, and M. Walter, *The mathematics of entanglement*, ArXiv e-prints, 2016.

[Bel64] J. S. Bell, *On the Einstein Podolsky Rosen paradox*, Phys. Phys. Fiz. **1** (1964), no. 3, 195–200, DOI 10.1103/PhysicsPhysiqueFizika.1.195. MR3790629

[Ben03] C. H. Bennett, *Notes on Landauer's principle, reversible computation, and Maxwell's demon: Quantum information and computation*, Stud. Hist. Philos. Sci. B Stud. Hist. Philos. Modern Phys. **34** (2003), no. 3, 501–510, DOI 10.1016/S1355-2198(03)00039-X. MR2011729

[Bir46] G. Birkhoff, *Three observations on linear algebra* (Spanish), Univ. Nac. Tucumán. Revista A. **5** (1946), 147–151. MR20547

[Bri87] M. Brion, *Sur l'image de l'application moment* (French), Séminaire d'algèbre Paul Dubreil et Marie-Paule Malliavin (Paris, 1986), Lecture Notes in Math., vol. 1296, Springer, Berlin, 1987, pp. 177–192, DOI 10.1007/BFb0078526. MR932055

[BS96] E. Bach and J. Shallit, *Algorithmic number theory. Vol. 1, Efficient algorithms*, Foundations of Computing Series, MIT Press, Cambridge, MA, 1996. MR1406794

[BW92] C. H. Bennett and S. J. Wiesner, *Communication via one- and two-particle operators on Einstein-Podolsky-Rosen states*, Phys. Rev. Lett. **69** (1992), no. 20, 2881–2884, DOI 10.1103/PhysRevLett.69.2881. MR1189330

[CcW18] M. Christandl, M. B. Şahinoğlu, and M. Walter, *Recoupling coefficients and quantum entropies*, Ann. Henri Poincaré **19** (2018), no. 2, 385–410, DOI 10.1007/s00023-017-0639-1. MR3748296

[CHM07] M. Christandl, A. W. Harrow, and G. Mitchison, *Nonzero Kronecker coefficients and what they tell us about spectra*, Comm. Math. Phys. **270** (2007), no. 3, 575–585, DOI 10.1007/s00220-006-0157-3. MR2276458

[Cho75] M. D. Choi, *Completely positive linear maps on complex matrices*, Linear Algebra Appl. **10** (1975), 285–290, DOI 10.1016/0024-3795(75)90075-0. MR376726

[CHSH69] J. F. Clauser, M. A. Horne, A. Shimony, and R. A. Holt, *Proposed experiment to test local hidden-variable theories*, Phys. Rev. Lett. **23** (1969), 880–884.

[Chu36] A. Church, *An unsolvable problem of elementary number theory*, Amer. J. Math. **58** (1936), no. 2, 345–363, DOI 10.2307/2371045. MR1507159

[CM06] M. Christandl and G. Mitchison, *The spectra of quantum states and the Kronecker coefficients of the symmetric group*, Comm. Math. Phys. **261** (2006), no. 3, 789–797, DOI 10.1007/s00220-005-1435-1. MR2197548

[Con] K. Conrad, *Cyclicity of* $(z/pz)^*$, available at `https://kconrad.math.uconn.edu/blurbs/grouptheory/cyclicmodp.pdf`.

[CT65] J. W. Cooley and J. W. Tukey, *An algorithm for the machine calculation of complex Fourier series*, Math. Comp. **19** (1965), 297–301, DOI 10.2307/2003354. MR178586

[CVZ18] M. Christandl, P. Vrana, and J. Zuiddam, *Universal points in the asymptotic spectrum of tensors*, STOC'18—Proceedings of the 50th Annual ACM SIGACT Symposium on Theory of Computing, ACM, New York, 2018, pp. 289–296, DOI 10.1145/3188745.3188766. MR3826254

[CVZ19] M. Christandl, P. Vrana, and J. Zuiddam, *Barriers for fast matrix multiplication from irreversibility*, 34th Computational Complexity Conference, LIPIcs. Leibniz Int. Proc. Inform., vol. 137, Schloss Dagstuhl. Leibniz-Zent. Inform., Wadern, 2019, pp. Art. No. 26, 17. MR3984631

[CW90] D. Coppersmith and S. Winograd, *Matrix multiplication via arithmetic progressions*, J. Symbolic Comput. **9** (1990), no. 3, 251–280, DOI 10.1016/S0747-7171(08)80013-2. MR1056627

[CW04] M. Christandl and A. Winter, *"Squashed entanglement": an additive entanglement measure*, J. Math. Phys. **45** (2004), no. 3, 829–840, DOI 10.1063/1.1643788. MR2036165

[DL42] G. C. Danielson and C. Lanczos, *Some improvements in practical Fourier analysis and their application to X-ray scattering from liquids*, J. Franklin Inst. **233** (1942), 365–380, 435–452, DOI 10.1016/S0016-0032(42)90767-1. MR6233

[DM35] M. Delbrülck and G. Molière, *Statistische quantenmechanik und thermodynamik*, Verlag der Akademie der Wissenschaften, Berlin, 1935.

[EPR35] A. Einstein, B. Podolsky, and N. Rosen, *Can quantum-mechanical description of physical reality be considered complete?*, Phys. Rev. **47** (1935), 777–780.

[Erd47] P. Erdös, *Some remarks on the theory of graphs*, Bull. Amer. Math. Soc. **53** (1947), 292–294, DOI 10.1090/S0002-9904-1947-08785-1. MR19911

[Fey85] R. Feynman, *QED*, Princeton University Press, 1985.

[Fey01] R. P. Feynman, *Feynman lectures on computation*, CRC Press, 2001.

[FH91] W. Fulton and J. Harris, *Representation theory: A first course; Readings in Mathematics*, Graduate Texts in Mathematics, vol. 129, Springer-Verlag, New York, 1991, DOI 10.1007/978-1-4612-0979-9. MR1153249

[FLS65] R. P. Feynman, R. B. Leighton, and M. Sands, *The Feynman lectures on physics. Vol. 3: Quantum mechanics*, Addison-Wesley Publishing Co., Inc., Reading, Mass.-London, 1965. MR213079

[FT76] E. S. Fry and R. C. Thompson, *Experimental test of local hidden-variable theories*, Phys. Rev. Lett. **37** (1976).

[Gau] C. F. Gauss, *Nachlass: Theoria interpolationis methodo nova tractata, Gauss, werke, band 3.*

[Geo21] I. Georgescu, *How the Bell tests changed quantum physics*, Nature Reviews Physics **3** (2021), 674–676.

[GHIL16] F. Gesmundo, J. D. Hauenstein, C. Ikenmeyer, and J. M. Landsberg, *Complexity of linear circuits and geometry*, Found. Comput. Math. **16** (2016), no. 3, 599–635, DOI 10.1007/s10208-015-9258-8. MR3494506

[Gle11] J. Gleick, *The information*, Pantheon Books, 2011.

[Gro96] L. K. Grover, *A fast quantum mechanical algorithm for database search*, Proceedings of the Twenty-eighth Annual ACM Symposium on the Theory of Computing (Philadelphia, PA, 1996), ACM, New York, 1996, pp. 212–219, DOI 10.1145/237814.237866. MR1427516

[GW09] R. Goodman and N. R. Wallach, *Symmetry, representations, and invariants*, Graduate Texts in Mathematics, vol. 255, Springer, Dordrecht, 2009, DOI 10.1007/978-0-387-79852-3. MR2522486

[Had67] D. Hadžiev, *Certain questions of the theory of vector invariants* (Russian), Mat. Sb. (N.S.) **72(114)** (1967), 420–435. MR223495

[Har01] L. Hardy, *Quantum theory from five reasonable axioms*, eprint `arXiv:quant-ph/0101012`, 2001.

[HD73] G. Hochschild and G. D. Mostow, *Unipotent groups in invariant theory*, Proc. Nat. Acad. Sci. U.S.A. **70** (1973), 646–648, DOI 10.1073/pnas.70.3.646. MR320174

[Hen21] N. Heninger, *RSA, DH and DSA in the wild*, Computational cryptography—algorithmic aspects of cryptology, London Math. Soc. Lecture Note Ser., vol. 469, Cambridge Univ. Press, Cambridge, 2021, pp. 140–181. MR4320717

[HHHH09] R. Horodecki, P. Horodecki, M. Horodecki, and K. Horodecki, *Quantum entanglement*, Rev. Modern Phys. **81** (2009), no. 2, 865–942, DOI 10.1103/RevModPhys.81.865. MR2515619

[HHL09] A. W. Harrow, A. Hassidim, and S. Lloyd, *Quantum algorithm for linear systems of equations*, Phys. Rev. Lett. **103** (2009), no. 15, 150502, 4, DOI 10.1103/PhysRevLett.103.150502. MR2551688

[HLP52] G. H. Hardy, J. E. Littlewood, and G. Pólya, *Inequalities*, 2nd ed., Cambridge, at the University Press, 1952. MR46395

[Hol] F. Holweck, *On the projective geometry of entanglement and contextuality*, Habilitation.

[HR00] G. H. Hardy and S. Ramanujan, *Asymptotic formulae in combinatory analysis [Proc. London Math. Soc. (2)* **17** *(1918), 75–115]*, Collected papers of Srinivasa Ramanujan, AMS Chelsea Publ., Providence, RI, 2000, pp. 276–309. MR2280879

[HvdH21] D. Harvey and J. van der Hoeven, *Integer multiplication in time $O(n \log n)$*, Ann. of Math. (2) **193** (2021), no. 2, 563–617, DOI 10.4007/annals.2021.193.2.4. MR4224716

[KEA+23] Y. Kim, A. Eddins, S. Anand, K. X. Wei, E. van den Berg, S. Rosenblatt, H. Nayfeh, Y. Wu, M. Zaletel, K. Temme, and A. Kandala, *Evidence for the utility of quantum computing before fault tolerance*, Nature **618** (2023), no. 7965, 500–505.

[Kle31] O. Klein, *Quantum coding*, Z. Phys. **72** (1931), 767–775.

[KLPSMN09] A. Kumar, S. V. Lokam, V. M. Patankar, and J. Sarma M. N., *Using elimination theory to construct rigid matrices*, Foundations of software technology and theoretical computer science—FSTTCS 2009, LIPIcs. Leibniz Int. Proc. Inform., vol. 4, Schloss Dagstuhl. Leibniz-Zent. Inform., Wadern, 2009, pp. 299–310. MR2870721

[Kly04] A. Klyachko, *Quantum marginal problem and representations of the symmetric group*, preprint arXiv:quant-ph/0409113v1, 2004.

[Kna02] A. W. Knapp, *Lie groups beyond an introduction*, 2nd ed., Progress in Mathematics, vol. 140, Birkhäuser Boston, Inc., Boston, MA, 2002. MR1920389

[KSB$^+$20] M. Kjaergaard, M. E. Schwartz, J. Braumüller, P. Krantz, J. I.-J. Wang, S. Gustavsson, and W. D. Oliver, *Superconducting qubits: Current state of play*, Annual Review of Condensed Matter Physics **11** (2020), no. 1, 369–395.

[KSV02] A. Yu. Kitaev, A. H. Shen, and M. N. Vyalyi, *Classical and quantum computation*, translated from the 1999 Russian original by Lester J. Senechal, Graduate Studies in Mathematics, vol. 47, American Mathematical Society, Providence, RI, 2002, DOI 10.1090/gsm/047. MR1907291

[KW01] M. Keyl and R. F. Werner, *Estimating the spectrum of a density operator*, Phys. Rev. A (3) **64** (2001), no. 5, 052311, 5, DOI 10.1103/PhysRevA.64.052311. MR1878924

[Lan61] R. Landauer, *Irreversibility and heat generation in the computing process*, IBM J. Res. Develop. **5** (1961), 183–191, DOI 10.1147/rd.53.0183. MR134833

[Lan12] J. M. Landsberg, *Tensors: geometry and applications*, Graduate Studies in Mathematics, vol. 128, American Mathematical Society, Providence, RI, 2012, DOI 10.1090/gsm/128. MR2865915

[Lan17] J. M. Landsberg, *Geometry and complexity theory*, Cambridge Studies in Advanced Mathematics, vol. 169, Cambridge University Press, Cambridge, 2017, DOI 10.1017/9781108183192. MR3729273

[Lan19] J. M. Landsberg, *Tensors: asymptotic geometry and developments 2016–2018*, CBMS Regional Conference Series in Mathematics, vol. 132, American Mathematical Society, Providence, RI, 2019, DOI 10.1090/cbms/132. MR3966415

[LG14] F. Le Gall, *Powers of tensors and fast matrix multiplication*, ISSAC 2014—Proceedings of the 39th International Symposium on Symbolic and Algebraic Computation, ACM, New York, 2014, pp. 296–303, DOI 10.1145/2608628.2608664. MR3239939

[Lie73] E. H. Lieb, *Convex trace functions and the Wigner-Yanase-Dyson conjecture*, Advances in Math. **11** (1973), 267–288, DOI 10.1016/0001-8708(73)90011-X. MR332080

[LR73] E. H. Lieb and M. B. Ruskai, *Proof of the strong subadditivity of quantum-mechanical entropy*, with an appendix by B. Simon, J. Mathematical Phys. **14** (1973), 1938–1941, DOI 10.1063/1.1666274. MR345558

[Lyo09] J. Lyons, *The house of wisdom*, Bloomsbury Press, 2009.

[Mac95] I. G. Macdonald, *Symmetric functions and Hall polynomials*, 2nd ed., with contributions by A. Zelevinsky, Oxford Science Publications, Oxford Mathematical Monographs, The Clarendon Press, Oxford University Press, New York, 1995. MR1354144

[Man15a] L. Manivel, *On the asymptotics of Kronecker coefficients*, J. Algebraic Combin. **42** (2015), no. 4, 999–1025, DOI 10.1007/s10801-015-0614-1. MR3417256

[Man15b] L. Manivel, *On the asymptotics of Kronecker coefficients, 2*, Sém. Lothar. Combin. **75** (2015), Art. B75d, 13. MR3461556

[Mir58] L. Mirsky, *Proofs of two theorems on doubly-stochastic matrices*, Proc. Amer. Math. Soc. **9** (1958), 371–374, DOI 10.2307/2032991. MR95180

[NC00] M. A. Nielsen and I. L. Chuang, *Quantum computation and quantum information*, Cambridge University Press, Cambridge, 2000. MR1796805

[Nie99] M. A. Nielsen, *Conditions for a class of entanglement transformations*, Physical Review Letters **83** (1999), 436–439.

[NLD+23] T. Nguyen, D. Lyakh, E. Dumitrescu, D. Clark, J. Larkin, and A. McCaskey, *Tensor network quantum virtual machine for simulating quantum circuits at exascale*, ACM Trans. Quantum Comput. **4** (2023), no. 1, Art. 6, 21, DOI 10.1145/3547334. MR4504805

[Plá] M. Plávala, *General probabilistic theories: an introduction*, Phys. Rep. **1033** (2023), 1–64, DOI 10.1016/j.physrep.2023.09.001. MR4642592

[Pre] J. Preskill, *Lecture notes*, `http://theory.caltech.edu/~preskill/ph219`.

[Pro07] C. Procesi, *Lie groups: An approach through invariants and representations*, Universitext, Springer, New York, 2007. MR2265844

[Rab80] M. O. Rabin, *Probabilistic algorithm for testing primality*, J. Number Theory **12** (1980), no. 1, 128–138, DOI 10.1016/0022-314X(80)90084-0. MR566880

[Res10] N. Ressayre, *Geometric invariant theory and the generalized eigenvalue problem*, Invent. Math. **180** (2010), no. 2, 389–441, DOI 10.1007/s00222-010-0233-3. MR2609246

[Res11] N. Ressayre, *Geometric invariant theory and generalized eigenvalue problem II* (English, with English and French summaries), Ann. Inst. Fourier (Grenoble) **61** (2011), no. 4, 1467–1491 (2012), DOI 10.5802/aif.2647. MR2951500

[RS+62] J. B. Rosser and L. Schoenfeld, *Approximate formulas for some functions of prime numbers*, Illinois J. Math. **6** (1962), 64–94. MR137689

[Rus04] M. B. Ruskai, *Lieb's simple proof of concavity of* Tr $A^pK^*B^{(1-p)}K$ *and remarks on related inequalities*, eprint `arXiv:quant-ph/0404126`, 2004.

[Sch95] B. Schumacher, *Quantum coding*, Phys. Rev. A (3) **51** (1995), no. 4, 2738–2747, DOI 10.1103/PhysRevA.51.2738. MR1328824

[Sch03] R. Schack, *Quantum theory from four of Hardy's axioms*, Special issue dedicated to David Mermin, Part I, Found. Phys. **33** (2003), no. 10, 1461–1468, DOI 10.1023/A:1026044329659. MR2039620

[Sha48] C. E. Shannon, *A mathematical theory of communication*, Bell System Tech. J. **27** (1948), 379–423, 623–656, DOI 10.1002/j.1538-7305.1948.tb01338.x. MR26286

[Sho94] P. W. Shor, *Algorithms for quantum computation: discrete logarithms and factoring*, 35th Annual Symposium on Foundations of Computer Science (Santa Fe, NM, 1994), IEEE Comput. Soc. Press, Los Alamitos, CA, 1994, pp. 124–134, DOI 10.1109/SFCS.1994.365700. MR1489242

[Sho97] P. W. Shor, *Polynomial-time algorithms for prime factorization and discrete logarithms on a quantum computer*, SIAM J. Comput. **26** (1997), no. 5, 1484–1509, DOI 10.1137/S0097539795293172. MR1471990

[Sim94] D. R. Simon, *On the power of quantum computation*, 35th Annual Symposium on Foundations of Computer Science (Santa Fe, NM, 1994), IEEE Comput. Soc. Press, Los Alamitos, CA, 1994, pp. 116–123, DOI 10.1109/SFCS.1994.365701. MR1489241

[SS71] A. Schönhage and V. Strassen, *Schnelle Multiplikation grosser Zahlen* (German, with English summary), Computing (Arch. Elektron. Rechnen) **7** (1971), 281–292, DOI 10.1007/bf02242355. MR292344

[STF+] Z.-Y. Shi, H. Tang, Z. Feng, Y. Wang, Z.-M. Li, J. Gao, Y.-J. Chang, T.-Y. Wang, J.-P. Dou, Z.-Y. Zhang, Z.-Q. Jiao, W.-H. Zhou, and X.-M. Jin, *Quantum fast hitting on glued trees mapped on a photonic chip*, Optica **7**, 613–618.

[Sti55] W. F. Stinespring, *Positive functions on C^*-algebras*, Proc. Amer. Math. Soc. **6** (1955), 211–216, DOI 10.2307/2032342. MR69403

[Sto] A. Stothers, *On the complexity of matrix multiplication*, PhD thesis, University of Edinburgh, 2010.

[Str69] V. Strassen, *Gaussian elimination is not optimal*, Numer. Math. **13** (1969), 354–356, DOI 10.1007/BF02165411. MR248973

[Val77] L. G. Valiant, *Graph-theoretic arguments in low-level complexity*, Mathematical foundations of computer science (Proc. Sixth Sympos., Tatranská Lomnica, 1977), Lecture Notes in Comput. Sci., Vol. 53, Springer, Berlin-New York, 1977, pp. 162–176. MR660702

[Val79] L. G. Valiant, *The complexity of computing the permanent*, Theoret. Comput. Sci. **8** (1979), no. 2, 189–201, DOI 10.1016/0304-3975(79)90044-6. MR526203

[VBHAF23] A. Vesperini, G. Bel-Hadj-Aissa, and R. Franzosi, *Entanglement and quantum correlation measures for quantum multipartite mixed states*, Scientific Reports **13** (2023), 2045–2322.

[VSD86] A. Vergis, K. Steiglitz, and B. Dickinson, *The complexity of analog computation*, Math. Comput. Simul. **28** (1986), no. 2, 91–113.

[VXXZ23] V. V. Williams, Y. Xu, Z. Xu, and R. Zhou, *New bounds for matrix multiplication: from alpha to omega*, Proceedings of the 2024 Annual ACM-SIAM Symposium on Discrete Algorithms (SODA), SIAM, Philadelphia, PA, 2024, pp. 3792–3835, DOI 10.1137/1.9781611977912.134. MR4699379

[vzGG13] J. von zur Gathen and J. Gerhard, *Modern computer algebra*, 3rd ed., Cambridge University Press, Cambridge, 2013, DOI 10.1017/CBO9781139856065. MR3087522

[Wat18] J. Watrous, *The theory of quantum information*, Cambridge University Press, 2018.

[Wil] V. Williams, *Breaking the Coppersimith-Winograd barrier*, preprint.

[Yur03] F. Yura, *Entanglement cost of three-level antisymmetric states*, J. Phys. A **36** (2003), no. 15, L237–L242, DOI 10.1088/0305-4470/36/15/104. MR1984497

Index

Selected Published Titles in This Series

243 **J. M. Landsberg,** Quantum Computation and Quantum Information, 2024
242 **Jayadev S. Athreya and Howard Masur,** Translation Surfaces, 2024
241 **Thorsten Theobald,** Real Algebraic Geometry and Optimization, 2024
238 **Julio González-Díaz, Ignacio García-Jurado, and M. Gloria Fiestras-Janeiro,** An Introductory Course on Mathematical Game Theory and Applications, Second Edition, 2023
237 **Michael Levitin, Dan Mangoubi, and Iosif Polterovich,** Topics in Spectral Geometry, 2023
235 **Bennett Chow,** Ricci Solitons in Low Dimensions, 2023
234 **Andrea Ferretti,** Homological Methods in Commutative Algebra, 2023
233 **Andrea Ferretti,** Commutative Algebra, 2023
232 **Harry Dym,** Linear Algebra in Action, Third Edition, 2023
231 **Luís Barreira and Yakov Pesin,** Introduction to Smooth Ergodic Theory, Second Edition, 2023
230 **Barbara Kaltenbacher and William Rundell,** Inverse Problems for Fractional Partial Differential Equations, 2023
229 **Giovanni Leoni,** A First Course in Fractional Sobolev Spaces, 2023
228 **Henk Bruin,** Topological and Ergodic Theory of Symbolic Dynamics, 2022
227 **William M. Goldman,** Geometric Structures on Manifolds, 2022
226 **Milivoje Lukić,** A First Course in Spectral Theory, 2022
225 **Jacob Bedrossian and Vlad Vicol,** The Mathematical Analysis of the Incompressible Euler and Navier-Stokes Equations, 2022
224 **Ben Krause,** Discrete Analogues in Harmonic Analysis, 2022
223 **Volodymyr Nekrashevych,** Groups and Topological Dynamics, 2022
222 **Michael Artin,** Algebraic Geometry, 2022
221 **David Damanik and Jake Fillman,** One-Dimensional Ergodic Schrödinger Operators, 2022
220 **Isaac Goldbring,** Ultrafilters Throughout Mathematics, 2022
219 **Michael Joswig,** Essentials of Tropical Combinatorics, 2021
218 **Riccardo Benedetti,** Lectures on Differential Topology, 2021
217 **Marius Crainic, Rui Loja Fernandes, and Ioan Mărcuţ,** Lectures on Poisson Geometry, 2021
216 **Brian Osserman,** A Concise Introduction to Algebraic Varieties, 2021
215 **Tai-Ping Liu,** Shock Waves, 2021
214 **Ioannis Karatzas and Constantinos Kardaras,** Portfolio Theory and Arbitrage, 2021
213 **Hung Vinh Tran,** Hamilton–Jacobi Equations, 2021
212 **Marcelo Viana and José M. Espinar,** Differential Equations, 2021
211 **Mateusz Michałek and Bernd Sturmfels,** Invitation to Nonlinear Algebra, 2021
210 **Bruce E. Sagan,** Combinatorics: The Art of Counting, 2020
209 **Jessica S. Purcell,** Hyperbolic Knot Theory, 2020
208 **Vicente Muñoz, Ángel González-Prieto, and Juan Ángel Rojo,** Geometry and Topology of Manifolds, 2020
207 **Dmitry N. Kozlov,** Organized Collapse: An Introduction to Discrete Morse Theory, 2020
206 **Ben Andrews, Bennett Chow, Christine Guenther, and Mat Langford,** Extrinsic Geometric Flows, 2020
205 **Mikhail Shubin,** Invitation to Partial Differential Equations, 2020

For a complete list of titles in this series, visit the
AMS Bookstore at **www.ams.org/bookstore/gsmseries/**.